Face Parts

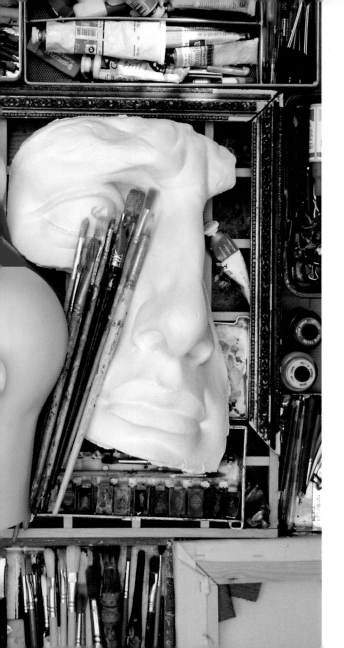

Face Parts

A visual source book for depicting the human face

Simon Jennings

MITCHELL BEAZLEY

Contents

The key to success in depicting the human face is to follow some very simple steps. Learn to look carefully, particularly when drawing from a model; casual glances gather little information and leave no memory. Always select what you feel is important from the immense amount of visual information in front of you, deciding how much to incorporate and whether you intend to simplify or abstract your creation or work it up with a lot of detail. This, in turn, will give you a clue as to how large the work should be and what media to select. Whatever form your final artwork takes it is often easier to start by making actual marks or brushstrokes on a sympathetic surface, usually paper. Making marks and lines, and applying areas of tone or washes — the physical act of drawing — initially demand a great deal of attention from an artist. However, looking and choosing are just as vital, and the marks you make will only fall into place and create a plausible image if they are grounded in good observation. Look at least as much as you draw, if not more! Finally, check your marks and make sure they really represent what you have seen and chosen. Too thin? Too dark? Wrong angle? Change them right away, because a successful artwork depends upon the relationships between the various elements, and to misrepresent one part will distort others. A keen critical faculty is required, too; frequently compare your work to your subject. If something does not look or feel right, change it immediately. The ability to appraise and revise your work as you go along is an essential skill to develop, and will contribute immensely to the success of your finished work.

behind the face

10 Foundations

The human skull has a whole range of functions beyond the job of mere protection, all of which influence its structure and pro- portions.

As an artist, you will find that a basic knowledge of the bones that underpin the physiognomy, and the muscles lying on top of them, will help you convey the character of any model. Although the skull determines the overall shape of the head, it is, in fact, the muscles that lie beneath the skin of the face, scalp, and neck that create the facial expressions that give movement and character to each individual.

While it is helpful to be aware of the many muscles responsible for lip movements, and those that surround the cavity of each eye, the problem confronting you as a portraitist is that all this underlying structure is hidden beneath layers of hair, skin, and fleshy features. You must rely on close observation to determine how much of what you know lies beneath the surface actually shows in the face you are drawing. The same attention to detail will enable you to record the huge range of emotions revealed by slight movements of facial muscles, and depict subtleties such as mood and temperament.

Expression
The eyes play an important role in these characteristic expressions, based on real-life observations from a 19th-century anatomical study of physiognomy.

Bones

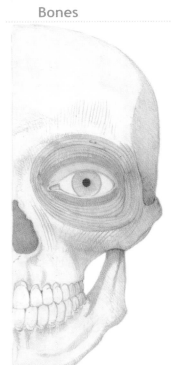

The skull determines the overall shape and proportion of the head.

Muscles

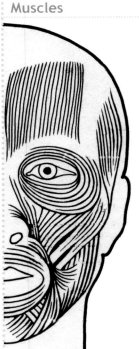

The muscles create facial expressions and movement.

Flesh

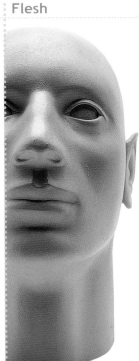

The skin is the covering for the underlying muscle and bone structure.

Features

Physiognomy means a person's facial features or expression. Skin, eyes, and hair bring character, texture, and colour to each individual.

12 | The human skull

The main features of the skull are the mandible, or lower jaw, and the cranium. The upper part of the cranium, called the calvaria, forms a box that encloses and protects the brain. The remainder of the skull forms the facial skeleton, of which the upper part is immovably fixed to the calvaria, and the lower part forms the freely movable mandible.

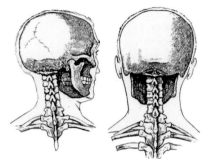

Mounting on the spine
The base of the skull rests on the vertebrae, forming the neck. The bones are subtly engineered to perform rocking and rotation movements.

• PARTS OF THE SKULL

• **Calvaria**
The rounded "cap" of the skull.
• **Cranium**
The whole skull, except for the jawbone.
• **Superciliary arch**
This bony ridge is visible just below the eyebrows. In some people, the entire eye socket is noticeable.
• **Zygomatic bone**
Also known as the cheekbone, this is most obvious in a half-profile view.
• **Mandible**
The movable lower jaw includes the chin.

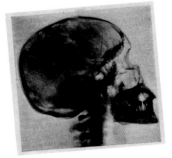

Early X-ray of the human skull

Look for the overall shape of the head; think about the structure of the skull and the proportions of its component parts.

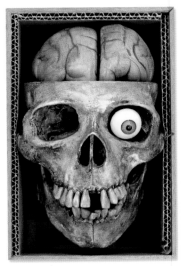

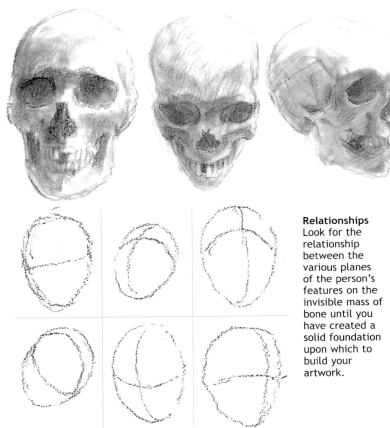

Relationships
Look for the relationship between the various planes of the person's features on the invisible mass of bone until you have created a solid foundation upon which to build your artwork.

14 | The muscles

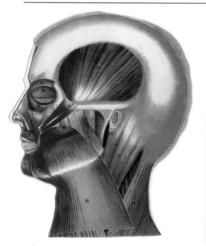

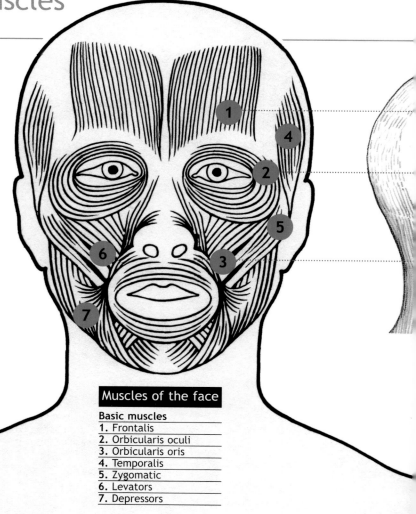

The bony skull determines the overall shape of the head. The muscles that lie between the skin and bone influence the basic facial contours and create the expressions that animate and give character to each individual.

Muscles of the face

Basic muscles
1. Frontalis
2. Orbicularis oculi
3. Orbicularis oris
4. Temporalis
5. Zygomatic
6. Levators
7. Depressors

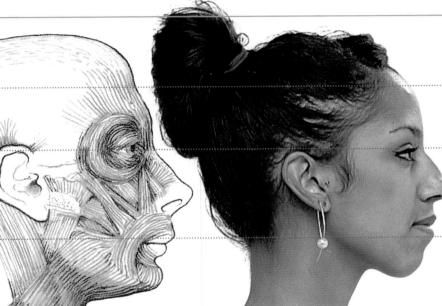

Frontalis
The muscle that controls the forehead and eyebrows.

Orbicularis oculi
These circular muscles open and close the eyelids.

Orbicularis oris
Another circular muscle, this opens and closes the mouth.

Orbital openings
The orbital openings in the skull contain not only the eyes, but all their associated muscles. With the head erect, the lower margin of these orbital openings and the upper margins of the external acoustic openings, or meatuses, of the ear, are on the same plane, known as the Frankfurt plane. Broadly speaking, this is halfway up the head, but it will not appear so if the head is tipped forward or back.

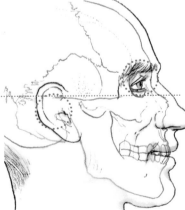

Veins of the face
Diagram showing the normally, invisible network of facial veins

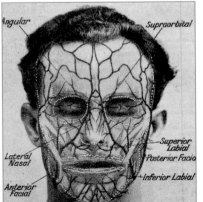

starting points

18 Thoughts on proportions

The head to body relationship

A convenient way of checking the relative proportions of a head to a figure is to take the length of the head as a measure. The number of "heads" in a figure varies, depending on age. In infancy, measuring from the crown of the skull to the soles of the feet, there are 4 heads in a figure; this rises to 4½ heads at the age of two. At the age of eight, there are about 6 heads; and at fourteen, nearly 7 heads. The figure of an adult person is equal to 7½ heads, shrinking to 7 in old age.

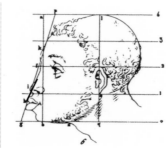

Strange proportions
"There is no excellent Beauty that hath not some Strangenesse in the Proportion."
Francis Bacon (1561-1626)

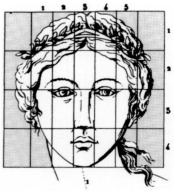

An Early Renaissance guide to facial proportions
"The face is divided into three parts, namely: the forehead, one; the nose, another; and from the nose to the chin, another. From the side of the nose through the whole length of the eye, one of these measures. From the end of the eye up to the ear, one of these measures. From one ear to the other, a face lengthwise... From the chin under the jaw to the base of the throat, one of the three measures. The throat, one measure long. From the pit of the throat to the top of the shoulder, one face."

From *Il Libro dell'Arte* by Cennino Cennini, c.1435

See also
Eyes: proportion and position, page 40
Features: proportion and position, page 56

Starting points
Basic proportions | 19

General proportions

The human face is rarely symmetrical. For example, eyes can vary in colour, size, and position; one ear may be different from the other; a mouth may be uneven. It is these combined differences which contribute to the unique character of an individual. However, there are certain "general, time-honoured" proportions that are helpful to know (see opposite page).

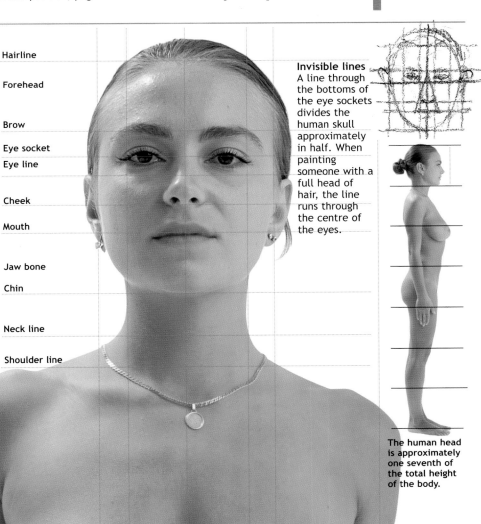

Hairline

Forehead

Brow

Eye socket

Eye line

Cheek

Mouth

Jaw bone

Chin

Neck line

Shoulder line

Invisible lines

A line through the bottoms of the eye sockets divides the human skull approximately in half. When painting someone with a full head of hair, the line runs through the centre of the eyes.

The human head is approximately one seventh of the total height of the body.

20 | Composition and format

Decide on your viewpoint (full face, profile, or three-quarter view), and whether you want to depict just the face, the head alone, the head and shoulders, or perhaps a half-length portrait that includes the clothed or nude torso. Much depends on how the model inspires you, and how you think you could best convey his or her unique character.

Composition

You might find that making a few preparatory sketches, or perhaps taking a few reference photographs, will help you begin to see what is particularly interesting about your subject.

Format

Part of the same process is to decide on the format. Although a vertical support is known as the "portrait" view, for some subjects the horizontal, "landscape", mode may give you a more successful result by providing scope for a better overall design.

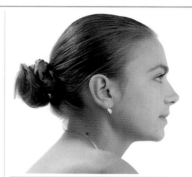

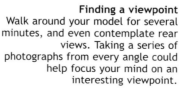

Finding a viewpoint
Walk around your model for several minutes, and even contemplate rear views. Taking a series of photographs from every angle could help focus your mind on an interesting viewpoint.

See also
Animated images,
pages 168-173

Posing your model
Before you settle on a pose, try out various views, from full face to three-quarter view and profile. Tilting the head up or down offers even more possibilities.

Head and shoulders
This painting is a conventional head-and-shoulders portrait against a plain background that emphasizes the contours of the face and throws the features into stark contrast. The artist has selected a three-quarter view, which makes it easier to see the features in three dimensions.

Nick Hyams
Joyce
Acrylic on paper
38 x 28cm (15 x 11in)

Head and torso
Here the artist has included more of the sitter in a three-quarter length portrait that focuses on the man's face. The clothing and torso are barely outlined, yet the hands, like the face, are drawn with sensitive detail, the inclusion of which add character and a sense of tranquillity to the overall study.

Victor Ambrus
Study
Graphite on tinted paper
41 x 28cm (16 x 11in)

22 Working with a model

See also
In the portrait studio, pages 76-103
Self-portraits, pages 108-123

There is no question that portraiture can be somewhat daunting, even for experienced artists. It can be difficult to concentrate on the mechanics of drawing and painting when, at the back of your mind, you are wondering what your sitter is going to think of your efforts.

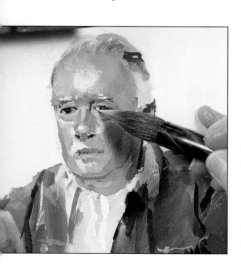

One way to avoid embarrassment is to paint a self-portrait (see pages 108-23). You can then be totally objective, and work as long as you like without fear of tiring your model!

Other obvious candidates are your friends and members of your family, who usually prove to be willing participants. Painting people you know intimately should help you bring out their true character, provided you can discipline yourself to look beyond your own preconceptions.

Whatever the relationship with your sitter, you are more likely to approach the task with enthusiasm if you genuinely feel that person will make an interesting subject to paint.

Lighting will affect your subject. A face under even illumination can appear almost flat, whereas a

strong sidelight throws the features into relief.

You could show more of your sitter, and include a setting that helps to convey the model's lifestyle or interests, but if your objective is to achieve a recognizable likeness, it pays to keep it simple.

Help your model to feel comfortable and relaxed – don't expect anyone to hold a smile or other exaggerated expression for any length of time. Be encouraging, and involve your sitter in the process by allowing him or her to see and comment on your work during the breaks.

Agree on what your sitter should wear. It does not have to be anything elaborate, but the addition of jewellery, a scarf or necktie, or even a hat, or some other prop may contribute something to the success of the painting.

First-hand knowledge
Portrait painter Valerie Wiffen – a contributing
artist to this book – is seen here at work on a
gouache study of the author while at the same
time she discusses the intricacies of working
with a live portrait model.

See also
Self-portrait sketchbook, pages 121-123
Visual reference, pages 178-179

A sketchbook can be a means of recording fleeting impressions, colour notes, and ideas for turning into paintings. It can be a portable scrapbook in which to collect interesting printed, pictorial references. It can be a notebook for observations and ideas that, one day, may provide that essential spark of inspiration. And, if nothing else, a sketchbook gives you somewhere to develop and practise your drawing and painting skills.

Printed ephemera
A sketchbook is an ideal portable scrapbook in which to collect a variety of pictorial references.

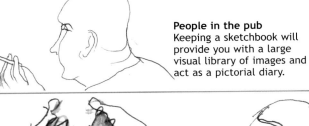

People in the pub
Keeping a sketchbook will provide you with a large visual library of images and act as a pictorial diary.

Everyday situations
Making sketches in everyday situations will improve your visual memory and sharpen your powers of observation. Because you do not have the luxury of spending much time on a single pose, your drawing speed should improve, and the more you sketch, the better your eye-to-hand coordination will become.

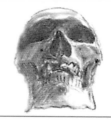

Some artists adopt a bold approach, attacking their work with gusto. Such works are often alive with energy and amply reward the risks taken by the artist. However, what appears to be carefree abandon is often the result of years of experience and experimentation.

If you are a beginner you may have a better chance of success if you adopt a systematic approach, remembering to step back from time to time to check and evaluate your work as it progresses. If it appears wrong, do not be afraid to modify or even obliterate it and start again.

Before you commit yourself to paint, it is worth making several studies and then a preliminary drawing of your subject. Drawing concentrates the mind and gives you an opportunity to observe your sitter closely.

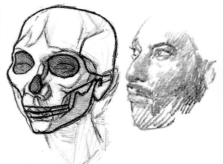

Underlying structure
A basic understanding of how the head is constructed (see pages 10-15) will help you to draw and paint with more confidence.

It is the underlying bone structure that gives us important clues to work with. The eyes are set deep within roughly circular sockets in the skull (you can detect them beneath the layers of skin and muscle). The cheek bones just below the sockets are invariably prominent features, as is the curve of the lower jawbone. Even when it is covered with hair, the size and shape of the cranial dome is vital to the overall proportion of the sitter's head.

Exercise in contours
It is easy to focus on the outline or edges of a shape, to the detriment of the form. Drawing contours helps you realize the three-dimensional nature of the head and face, and makes you look at the overall form.

Make visual notes and basic sketches to familiarize yourself with the subject.

Starting points are as varied as your subjects, and will be different every time you begin to draw. Perhaps it will be the angle of a nose, the shape of a shadow across a cheek, a diagonal strand of hair - any of these could capture your imagination and help to get you going.

However, resist the temptation to concentrate on details at the expense of the overview. If in doubt, always go for the larger option, like the overall shape of the head. Look for the relationship between the various planes of the person's features until you have created a solid foundation (see Exercise in contours, *left*).

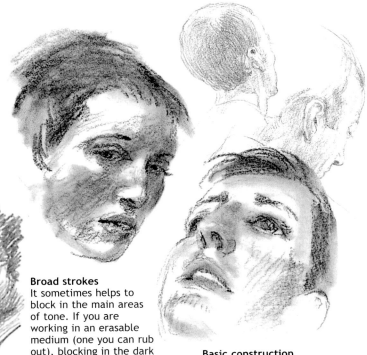

Broad strokes
It sometimes helps to block in the main areas of tone. If you are working in an erasable medium (one you can rub out), blocking in the dark or shadow tones as you go along enables you to get a feel for the three-dimensional nature of what you are drawing.

Basic construction
Solve any problems of perspective or proportion before going for full-strength marks and specific details. Once you have achieved a sound framework, the details will fall into place naturally.

The choice of materials can dictate whether a drawing will take a linear or tonal direction.

Line drawings
A fine-nibbed drawing pen, sometimes known as a mapping pen, was used to draw these simple linear profiles (*below*).

Experiment with line
Pen and ink is the ideal medium for line drawing. Ordinary drawing pens have flexible steel nibs that produce a variety of line widths and marks, depending on the pressure applied. Technical pens such as those shown below produce a line of unvarying width regardless of pressure or the direction in which the pen is moved.

Line and stipple
The drawing on the left was made with a technical pen containing sepia ink. Tone is implied with a dot-stipple technique and simple hatching.

Mapping pen drawings

Experimenting with tone
Drawing with a brush produces a lively,
fluid style and the ability to impart
tone into a drawing.

Graphite pencil
A soft pencil is used to provide the linear structure of
this drawing, while soft shading on the rough texture of
the paper creates tonal depth and shadows.

Brush
drawing

Soft-graphite
pencil on rough
paper

Charcoal
Charcoal is the classic artist's
medium for drawing. It is
ideal for rubbing and blending
for a variety of tonal effects.

Experiment with washes and colour to give drawings volume, density, and the illusion of light and shade. Don't be afraid to experiment with brushes using pure, direct colour to extend your range.

Washes
Create tone and areas of deep shadow in your drawings (*left* and *below*) by applying transparent washes over initial linework.

Line and wash drawings
Using monochrome and colour washes enables you to define features with colour and tone rather than linework. It is a very direct and fast medium. Start by applying pale washes to your basic linework drawing, then overlay them with stronger tone – as a wash sinks into the paper it creates an area of tone that may be impossible to modify, so err on the side of caution until you have built a firm foundation for your drawing.

See also
Colour and tone, page 75
Portrait colours, pages 174-175

Starting points
Pure colour | 31

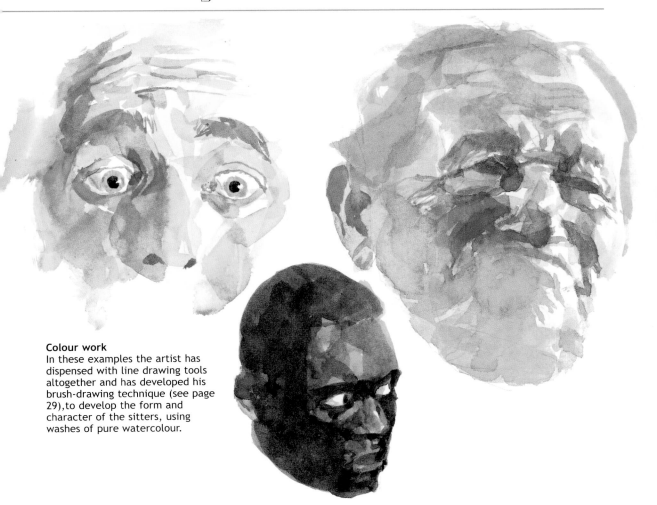

Colour work
In these examples the artist has dispensed with line drawing tools altogether and has developed his brush-drawing technique (see page 29), to develop the form and character of the sitters, using washes of pure watercolour.

Look especially for the fall of light across the head and surfaces of the face. Depicting this play of light and shade will give you a convincingly solid and realistic image of the person.

Building up tones
Carefully observe light and shade for a solid and three-dimensional image, and try building up tones gradually. Here (*left*) the artist drew herself, using three grades of charcoal pencil in three-quarter shadow. The dramatic lighting serves to accentuate the form and volume of the portrait.

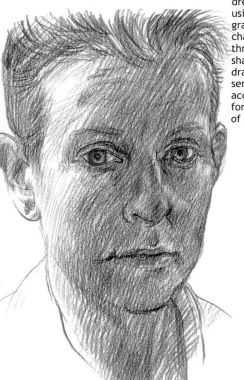

White on black
Drawing with white pencil on dark paper is a good way to become aware of light. Instead of drawing marks that really describe three dimensions, start thinking you are drawing with light. With this exercise, you have to draw the highlights not the shadows – a complete reversal of standard practice.

Tonal design

Tone describes the relative lightness or darkness of a colour; it is these light and dark areas that make the tonal pattern of a portrait. This pattern of light and dark across the face attracts attention and influences the mood and content of a portrait. Tone is easier to understand when dealing with black, white, and shades of grey, but all colours exhibit a range of tones – put simply, the more a surface turns away from the light the darker it becomes.

Light sources

For dramatic effect try to ensure that illumination comes from one source only. Side lighting emphasizes form and facial structure.

features and details
eyes

Seeing
A section of the human head by Leonardo da Vinci, c.1489-92, speculates on the connection of the eye to the brain.

In the days when the study of anatomy was still in its infancy, artists like Leonardo da Vinci filled pages of notebooks with detailed drawings and sketches of eyes, from bare spheroid eyeballs to the enigmatic, coded gaze that eventually graced the Mona Lisa. You do not have to go to such obsessive lengths, but a working knowledge of the anatomy of the eye will undoubtedly improve your powers of observation, and, from that, your art.

It pays to develop a broad understanding of how an eye fits in the skull and how its appearance is affected by the muscles and skin that surround it. The individual components – the eyeball, socket, muscles, eyelids, eyebrows, and lashes – appear to be very simple, but the interplay and relationship between these various components can be extraordinarily subtle, making the human eye capable of a wide range of movement that, in turn, conveys a whole gamut of emotions. The eyes are not called the "mirrors of the soul" for nothing – no other feature can express so much with such little variation.

And there is little or no room for error when trying to convey these subtleties in a drawing. We are so used to reading the messages sent out by eyes, that even the slightest misplacement of what seems an insignificant line can completely change our reading of someone's portrait.

When drawing a model or a self-portrait, notice how the smallest shift in position will alter the whole mood of the eyes; understanding why this is so, and how the eyes move and are controlled, goes a long way towards being able to convey mood and character accurately.

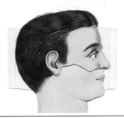

Teaching aid
(early 20th century)
By lifting the
various flaps and
covers of this
ingenious piece of
printed-paper
engineering the
anatomy of the eye
gradually unfolds.

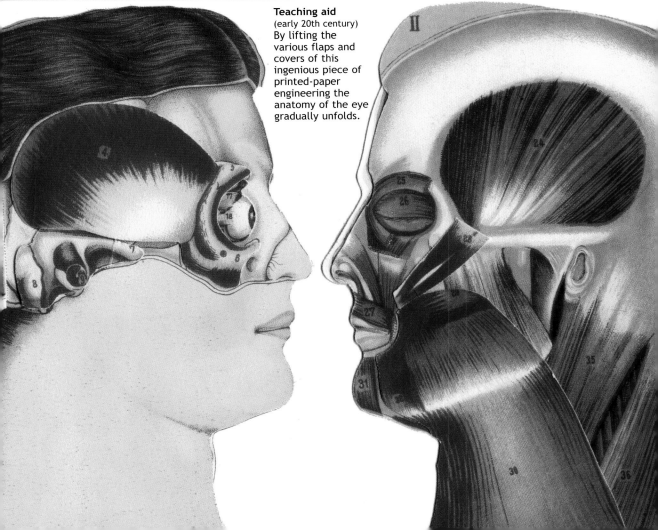

When drawing an eye, remember that it is not a flat image pasted to the front of your face. It is a three-dimensional object and is set within a hollow eye socket.

Tear duct, or lacrimal caruncle, in the corner of the eye

Iris

Pupil

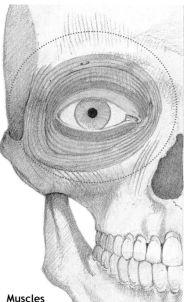

Muscles
The whole eye area is covered by a single, thin, circular sheet of muscle that closely follows the form of the skull. The muscles that rotate the eyeball are too deep to be seen by the artist; the bones and skin, and their relationship to the eyeball, are the more significant factors. However, what the muscles do to the eye and its surrounding tissue is very important, especially in terms of facial expressions.

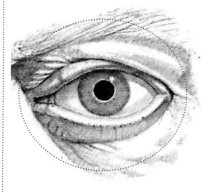

Eyeball
The eyeball is a moist sphere, with its own shadows and highlights. Observe carefully the relationship between the eyeball, lids, eyebrow, and muscles around the eye. Also note the little bump where the tear duct opens. The tear duct has shadows and highlights, too.

There are no hard and fast rules on how to draw eyes, but these pages of observations will help you to see the components of the human eye more clearly.

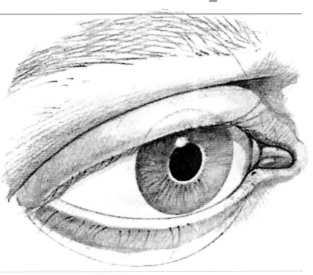

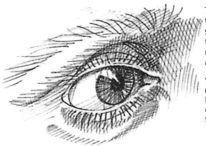

The brow ridge
The outer edge of the ridge is relatively sharp, but it tends to thicken and become rounded as it approaches the nose.

The lower ridge
On some people, the lower ridge of the eye socket hardly shows.

Spheres
The human eyeball is approximately spherical, but only a small part of the sphere is visible.

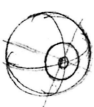

Positioning the pupil and iris
Use lines intersecting at right angles to mark the position of the pupil and iris.

Eyelids
The eyelids follow the curvature of the eyeball and vary in shape from person to person, and with changing facial expressions.

Eyebrows
Note the direction of growth and how the hairs tend to follow the line of the bony ridge beneath.

Cornea
The cornea is a protective membrane that projects slightly, forming the front surface of the eyeball. The iris and pupil are behind the transparent cornea. The bulging cornea can sometimes be seen clearly under the closed eyelid.

40 | Eyes: proportion and position

In an adult, the eyes are at the centre of the face, and the nose is longer in proportion. A baby's eyes are well below the centre of the face, and its nose is relatively short.

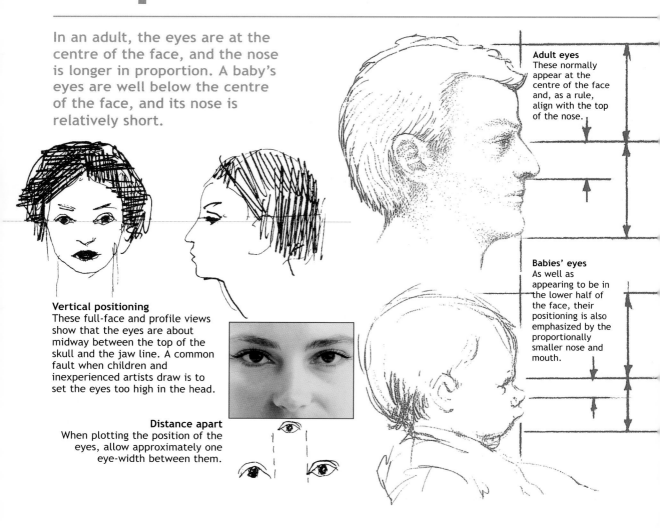

Adult eyes
These normally appear at the centre of the face and, as a rule, align with the top of the nose.

Babies' eyes
As well as appearing to be in the lower half of the face, their positioning is also emphasized by the proportionally smaller nose and mouth.

Vertical positioning
These full-face and profile views show that the eyes are about midway between the top of the skull and the jaw line. A common fault when children and inexperienced artists draw is to set the eyes too high in the head.

Distance apart
When plotting the position of the eyes, allow approximately one eye-width between them.

The iris is fully developed from birth and never becomes any bigger. As a result, the eyes of a baby or young child appear to be proportionally larger than those of an adult.

Wide-eyed innocence
You can see in these images how the eyes seem to dominate portraits of babies and young children. Proportionally larger than adult eyes, children's eyes are also bright and colourful, and surrounded with fresh, smooth skin. Close-up observation and studies reveal that children often have exceptionally long eyelashes that emphasize the eyes still further.

Relative sizes
Like the rest of our anatomy, eyes change a great deal as we get older. For an artist, it is both a pleasure and a challenge to depict the extremes of youth.

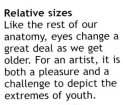

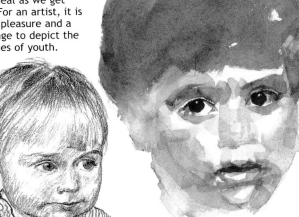

In older people, and those with deeply set eyes, you can see more clearly the entire rim of the socket beneath the skin. This becomes particularly evident as people advance in years.

Stretched skin
With increasing age, the skin stretches and sags, almost covering the upper eyelid. Note, also, the thickness of both eyelids.

Raised eyebrow
When the eyebrow is raised in surprise or delight, it appears to travel up the forehead, emphasizing the shape of the upper bony ridge of the eye socket.

Bony ridge
The bony ridge of the eye socket varies from person to person. It often casts deep shadows and tends to be particularly noticeable in older people.

The lower ridge
On some people, the lower ridge of the eye socket hardly shows externally, especially when pouches or "bags" form below the eye.

As a person ages, eyebrow hair often grows longer and more unruly. Individual eyebrow hairs that are longer than eyelashes tend to grow upwards and outwards.

Shape and direction
The shape of the eyebrow and the direction of the eyebrow hair help to describe the underlying structure of the eye socket.

Eyebrows and expression
Straight, upturned, and beetling eyebrows create very different expressions.

Random spacing
Eyelashes are somewhat randomly spaced along the lid, and often tend to clump together.

Eyebrows
The upper edges of the eye sockets are marked by the eyebrows; note how the socket edges actually pass across the line of the eyebrows, which appear darker at that point. Eyebrow hair also tends to grow vertically at the nearest point to the nose.

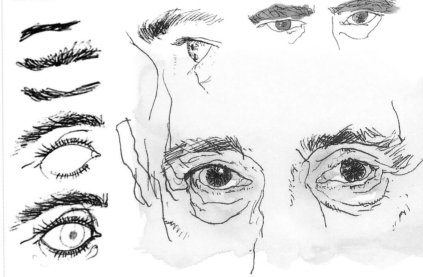

Eyelashes
Eyelashes mark the edge of the eyelids. They provide interesting textural detail, and draw attention to the eyes themselves. Long, dark eyelashes are generally considered attractive, and are often emphasized with makeup.

Drawing eyelashes
The lashes curve away from the eye, and usually taper sharply. Those on the outer edge of the upper lid tend to angle away from the nose and are normally longer than the others.

Expressions of emotion
Smiling, frowning, and other expressions of emotion, wrinkle the skin surrounding the eye, and alter the relationship of the eyelids.

44 | Eyes: colour and reflection

The colour of irises varies from person to person, and the so-called "whites of the eyes" are seldom, if ever, pure white – hold up a white card to someone's face to see the difference.

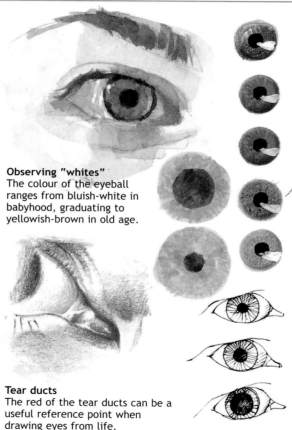

Observing "whites"
The colour of the eyeball ranges from bluish-white in babyhood, graduating to yellowish-brown in old age.

Tear ducts
The red of the tear ducts can be a useful reference point when drawing eyes from life.

Iris colour
This is never a flat tint, but is subtly and individually modulated; it is often darker on the outer edge. The iris is textured, with ridges and valleys.

Dilation and contraction
The size of the pupil affects the apparent colour of the eyes. The pupil changes in size, depending on light levels. When dilated, due to a low level of light, the pupil is large and the area of the coloured iris is correspondingly small. In bright light, the converse is true.

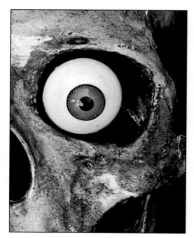

The "whites"
They are never pure, stark white as seen in this anatomical model.

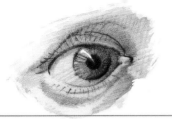

The smooth, wet surface of the eye produces mirror-like reflections, but because it is convex the reflected images are distorted. Shadows cast by the eyelids help describe the spherical shape of the eyeball.

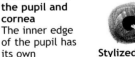

Highlights
Whether large or small, highlights and reflections will follow the contours of the surface of the eyeball.

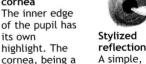

To show the curvature
Shade the eyeball lightly, so that it still appears to be white, and draw in the shadows cast by the upper lid. The coloured iris is darker, and you can hatch it with radial lines. The pupil, being a hole, is quite black.

Reflections in the pupil and cornea
The inner edge of the pupil has its own highlight. The cornea, being a transparent bubble protruding from the surface of the eye, displays reflections similar to the spherical eyeball.

Stylized reflection
A simple, stylized reflection drawn in the form of a curved wedge across the iris makes the eye look shiny and helps define the shape.

46 | Eyes: angles and rotation

The appearance of human eyes changes when they open and close, or when they rotate up or downwards, and from side to side. The shape of the eye also changes considerably, depending on the viewpoint from which it is being drawn.

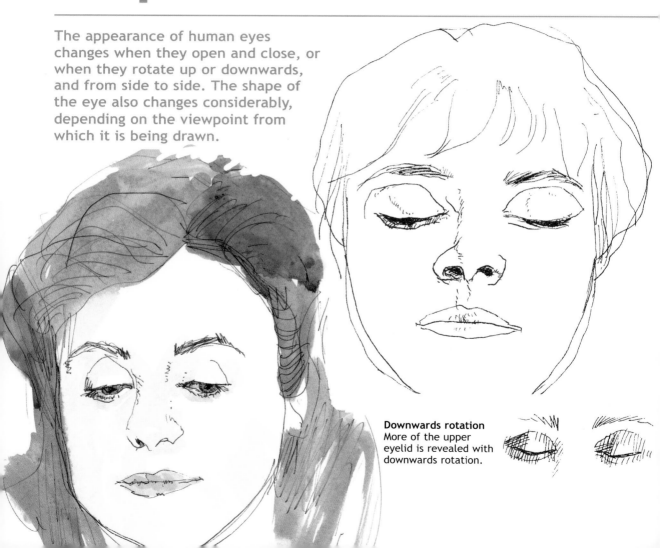

Downwards rotation
More of the upper eyelid is revealed with downwards rotation.

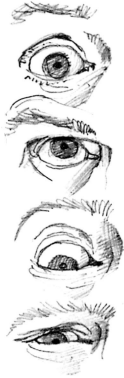

Eyes in profile
Whether the eyes are open or closed, the distinct, spherical nature of the eyeball is evident even when drawing portraits in profile. When the eye is closed, the lid emphasizes the spherical nature of the eye.

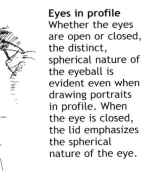

Rotation and movement
Moving the eyeball causes changes in the eyelids and the surrounding skin.

Slightly different
When a pair of eyes are turned aside, they may no longer look exactly the same, simply because one has to turn a little more than the other in order to focus on a distant object.

48 | Eyes and glasses

Spectacles can be used as props in many ways, and can bring a sense of theatre and drama to enliven the character of drawings of faces.

Looking over
A somewhat superior expression is achieved by the subject looking at the viewer over the top of his glasses.

Off the nose
The glasses are pulled down, which dramatically focuses attention on the subject's eyes.

By association, the wearing of sunglasses looks somewhat sinister. Apart from preventing eye-to-eye contact, sunglasses can resemble the dark, empty, eye sockets of a skull.

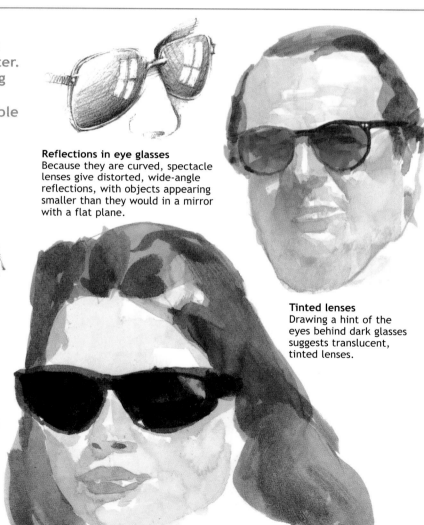

Reflections in eye glasses
Because they are curved, spectacle lenses give distorted, wide-angle reflections, with objects appearing smaller than they would in a mirror with a flat plane.

Hiding the eyes
Very dark sunglasses are an effective way of drawing attention to the eyes; because the eyes are invisible, we cannot be sure who or what the person is looking at, and the person's emotions or mood are a blank page.

Tinted lenses
Drawing a hint of the eyes behind dark glasses suggests translucent, tinted lenses.

See also
Line and tone, page 28
Line and wash, page 30
Glossary of materials, pages 180-181

Features and details
Studies of eyes | 51

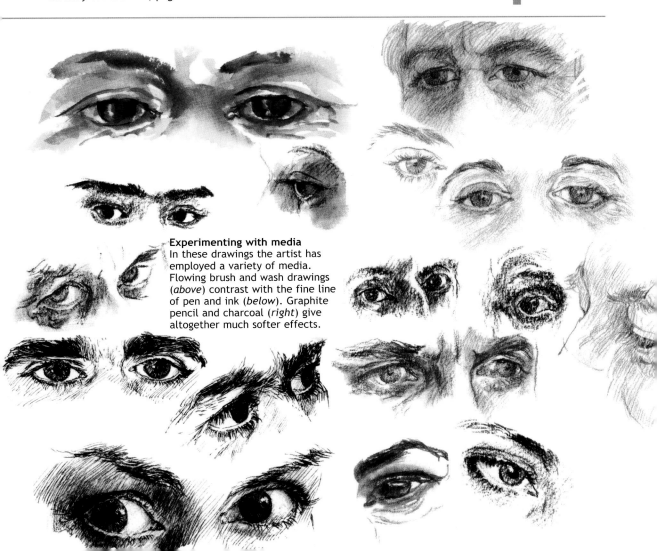

Experimenting with media
In these drawings the artist has
employed a variety of media.
Flowing brush and wash drawings
(*above*) contrast with the fine line
of pen and ink (*below*). Graphite
pencil and charcoal (*right*) give
altogether much softer effects.

Except for the colour of the iris, the eye itself is a fairly standard organ throughout the human species. It is the context of the eye – folds of skin, eyelashes, and eyebrows – that gives each pair of eyes their unique and personal character.

Eyebrows and lids
With some people, there is practically no gap between eyelids and eyebrows.

Drooping eyelids
The slack skin that partially covers the eyelids is usually a sign of ageing.

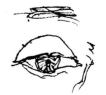

Large eyelids
Larger-than-average eyelids give the eyes a languid appearance.

Invisible eyelids
The eyelids are sometimes almost invisible, especially when squinting into bright light.

Eyelids
The outer limits of the upper eyelid often tend to overlap the lower lid. The upper lid is always the larger of the two and has a much greater range of movement. Some eyes have a distinctive fold of skin on the insides of the top lids.

See also
Older eyes, page 42
Expressive eyes, page 72

Features and details
Character 53

As people get older, the folds of the upper eyelids droop and rest on the eyelashes, pouches develop below the eyes, the radiating lines at the corners of the eyes deepen, and the frown lines between the eyebrows become permanent.

Lines

Whether you call them "crow's feet" or "laughter lines", the creases at the sides of the eyes are a great indication of character and personality, and imply mood and expression.

Sunken eyes

Eyes change gradually over the years, paralleling the changes in the whole face. In a craggy, sculptural face, the eyes can hardly be seen, and are merely dark holes under heavy brows.

Folds and creases

As many people age, their eyes seem to shrink into the increasingly complex folds and creases around them, and the iris is less prominent and paler in colour. Sometimes, the sockets become clearly defined as the eyes retreat into them.

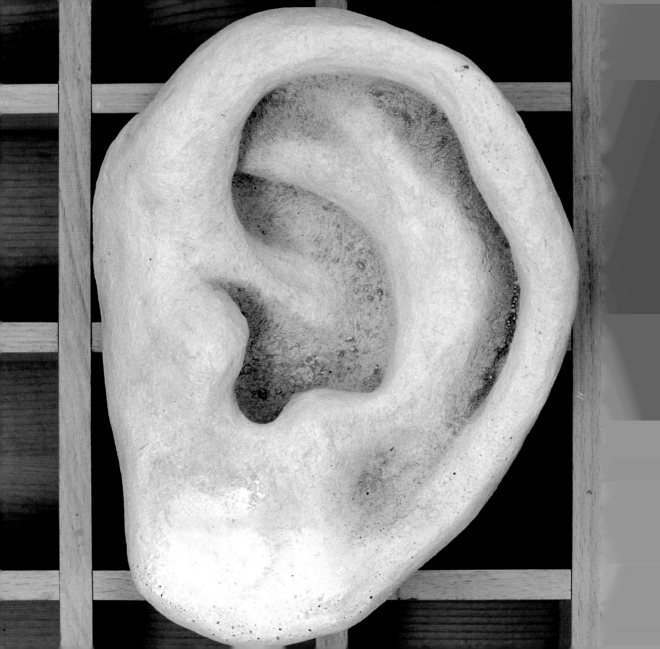

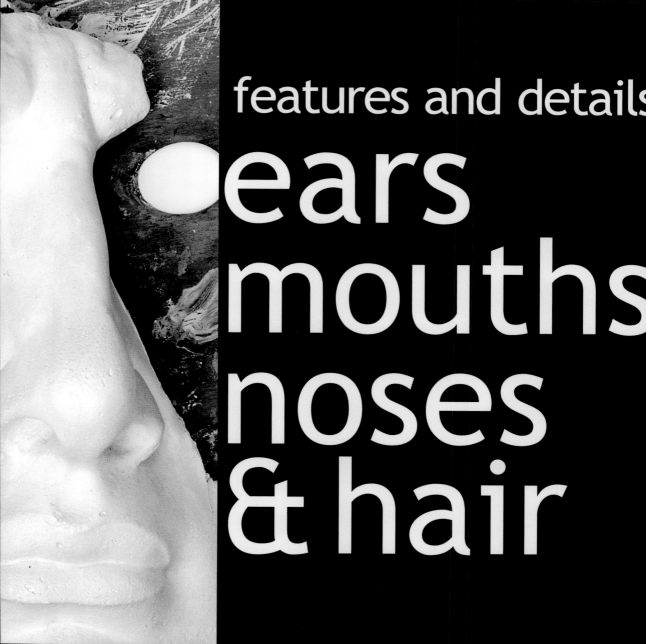

features and details

ears
mouths
noses
& hair

56 Features: proportion and position

We have seen on the previous pages something of the visual nature of the human eye. Similarly, in attempting an accurate depiction of the human face a knowledge of the form and structure of the other main facial features — the ears, nose, mouth, and hair — will go a long way towards obtaining a recognizable likeness.

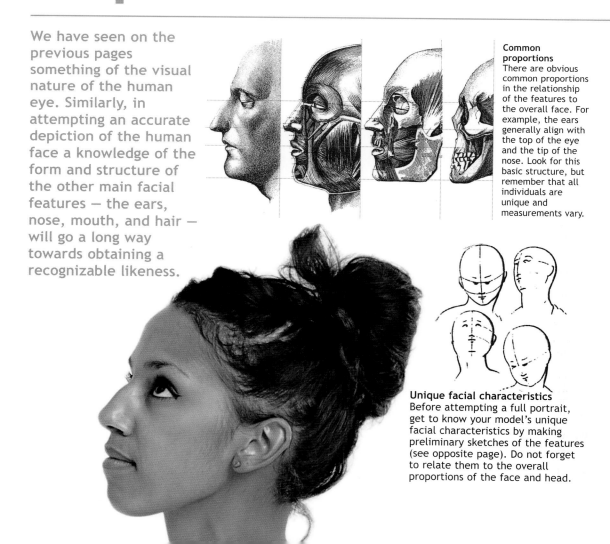

Common proportions
There are obvious common proportions in the relationship of the features to the overall face. For example, the ears generally align with the top of the eye and the tip of the nose. Look for this basic structure, but remember that all individuals are unique and measurements vary.

Unique facial characteristics
Before attempting a full portrait, get to know your model's unique facial characteristics by making preliminary sketches of the features (see opposite page). Do not forget to relate them to the overall proportions of the face and head.

See also
Basic proportions, page 19
Sketchbooks, page 25
Eyes: proportion and position, page 40

Techniques for preliminary sketches

In addition to portraying form, it is important to bring the volume and substance of the features into your drawing. Try to create the effect of light and shade by experimenting with shadow and shading techniques, either individually or in combination.

Hatching

Shade an area by drawing parallel lines close together.

Crosshatching

Overdraw hatched lines at an angle, to create a darker tone.

Shading

Create tone by shading with a soft pencil, or try a combination of the above techniques.

Solid shadows

Create areas of shadow with solid tone. Under certain lighting conditions, you will be unable to see highlights in the eye because of deep shadow cast by the eyelid and brow.

Dot and stipple

Create subtly modulated areas of tone by stippling with the point of a pen; the closer the dots, the darker the tone.

Creating volume

A strong side light can emphasize volume, increasing the dramatic intensity of the features.

Experiment

Do not be afraid to experiment with brush drawing and colour or tonal washes to give drawings greater volume and more density.

58 | Features: naming parts

As an artist you will probably never have to learn the anatomical names of body parts, but it is interesting to know that the little bump seen on most people's upper ear rim is called *"Darwin's tubercle"* and the two parts of the lower lip are known as the *"tori"*, from Latin *torus*, meaning cushion, mound or swelling.

Shell like
The ear is made of flexible tissue known as cartilage, and can be notoriously difficult to draw. Once you get the hang of it and realize that the ear is basically a simple spiral construction, not dissimilar to certain types of sea or snail shell (hence the term "shell like" used to describe it), the task of accurately depicting it becomes a lot easier.

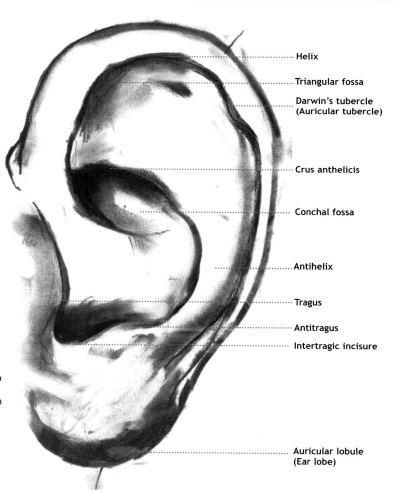

Helix

Triangular fossa

Darwin's tubercle
(Auricular tubercle)

Crus anthelicis

Conchal fossa

Antihelix

Tragus

Antitragus

Intertragic incisure

Auricular lobule
(Ear lobe)

See also
Behind the face, pages 8-15
Recording features and details,
pages 62-63

Features and details
Ears, mouths, noses | 59

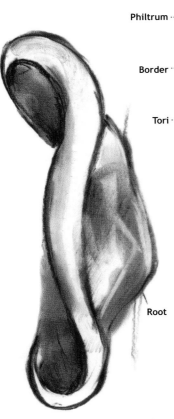

Ear, from the back

Philtrum

Border

Tori

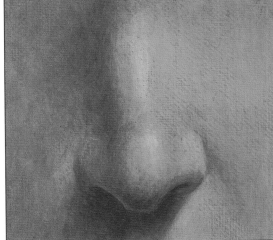

Root

George Underwood
Portrait Fragments
(*left and below left*)
Oil on canvas
10.2 x 10.2cm (4 x 4in)

60 Ears

Because ears are so complex, drawing them is often one of the more difficult aspects of portraiture. Each person's ears are so distinctive, they demand careful consideration if you want to achieve a convincing likeness.

Individuality
No two ears are exactly alike, even on the same head. Before you draw someone's finished portrait, it may help to make preliminary studies of their ears. Are they prominent, neatly tucked in, small, or large? Note that a pair of ears are rarely on the exact same level.

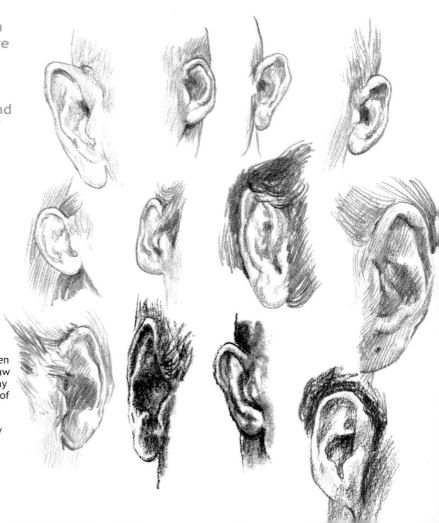

See also
Features: naming parts 58-59
Recording features and details, pages 62-63
Hair and ears, page 68

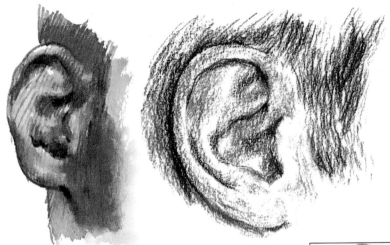

Exploring tonality
Using a medium such as wash (*far left*), charcoal pencil (*left*), or oil pastel (*below left*), gives you the opportunity to exploit tonality and to suggest depth, particularly on textured paper. An early 20th-century medical drawing (*below right*) reveals the complex inner workings of the ear.

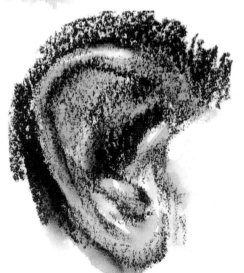

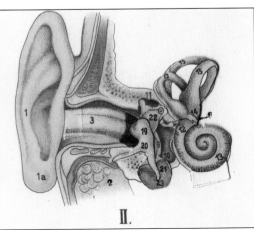

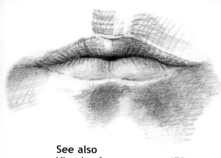

See also
Visual reference, page 178
Model reference, pages 104-105
Photographic sources, page 179

Recording features and details

Photographs of the features can often assist an artist in understanding the sweeping contours and the numerous planes of ears, mouths, and noses. The construction of an ear, as observed in the images on the right, can be seen very clearly in a photograph and the mouth's mobility and subtlety of expression, which is often elusive, can be helpfully frozen in time. Capturing the nose in two dimensions can help when attempting to visualize its protruding perspective and the delicately complex swell of the nostrils.

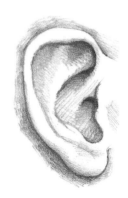

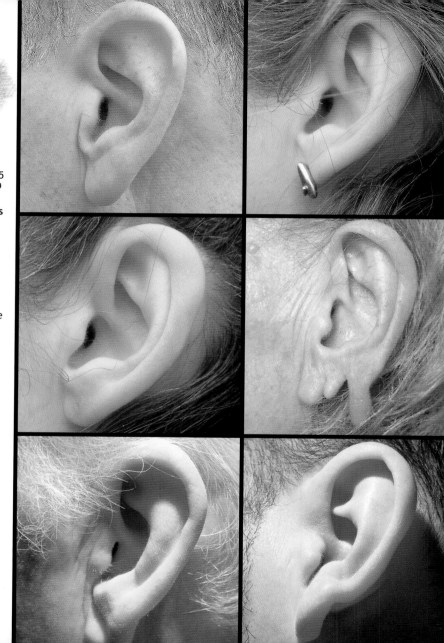

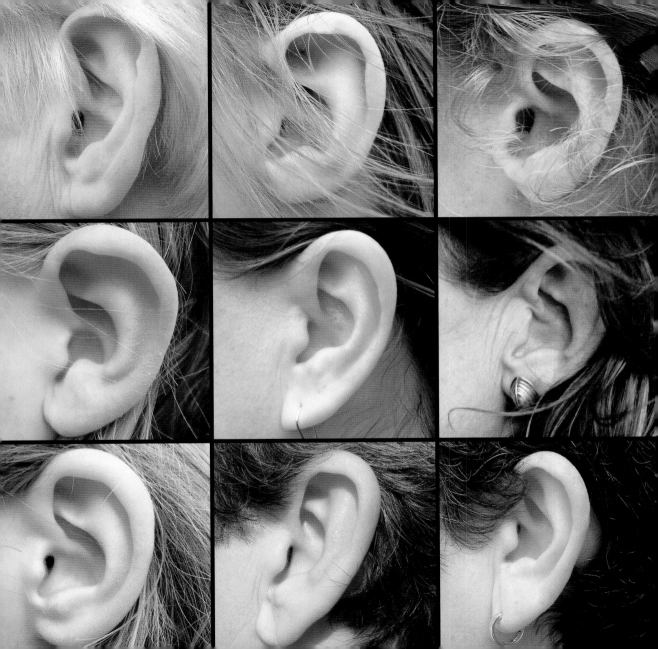

The mouth is amazingly expressive, and is a remarkable vehicle for conveying the model's character. An accurate portrayal of the mouth, within the overall shape of the face, is a key factor in catching a sitter's likeness.

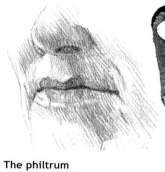

The mouth is extraordinarily expressive.

Feathered Head
(Hawaiian Islands, late 18th-century)
Made of feathers and basketwork, the head displays an expressive mouth of canine teeth and shells.
British Museum, London

The philtrum
Also known as the infranasal depression, the philtrum is the vertical groove in the upper lip. It allows humans to express a much larger range of lip motions than would otherwise be possible.

Because I sing
Alain Platel and The Shout, London, 2003
Publicity photographs for a theatrical performance show a wide variety of oral movement and expression.

Symmetry
Do not assume that a person's lips are perfectly regular. The asymmetrical or lopsided nature of some mouths is both fascinating and a challenge to draw.

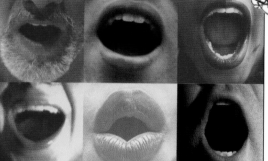

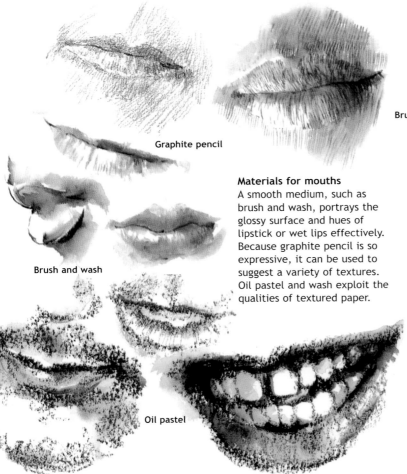

Graphite pencil

Brush and wash

Brush and wash

Materials for mouths

A smooth medium, such as brush and wash, portrays the glossy surface and hues of lipstick or wet lips effectively. Because graphite pencil is so expressive, it can be used to suggest a variety of textures. Oil pastel and wash exploit the qualities of textured paper.

Oil pastel

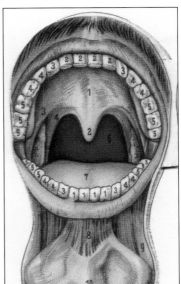

Inner workings

An early 20th-century medical drawing reveals the teeth and inner workings of the mouth.

66 Noses

Experiment with materials, techniques, and viewpoints to show how a nose stands out from the face. Here (and opposite) we see a variety of techniques in use: watercolour wash, charcoal, bold pastel, and fine pencil line.

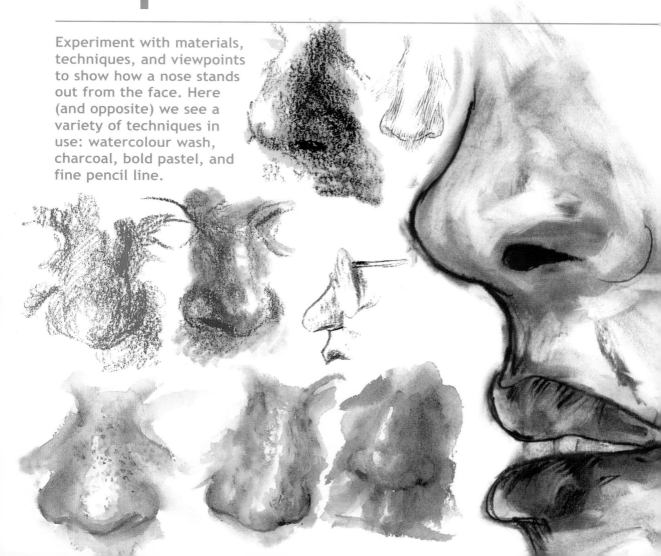

Marks, contours, and highlights

Linear marks can be used with precision to describe subtle contours.

The highlights along the ridge of the nose and above the nostrils give the nose a three-dimensional quality. The dark cavity of each nostril, and the deep-set crease on each side of the nose, accentuate the effect.

The shape of a person's nose is not directly related to the skull, and there is considerable variation in nose size, shape, and skin texture.

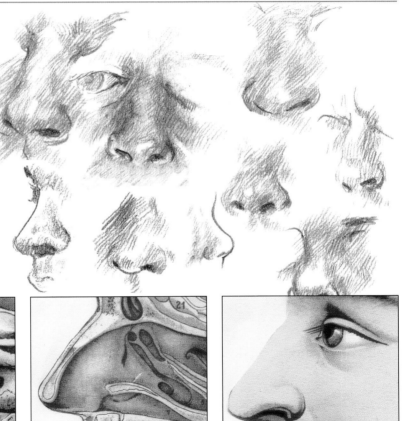

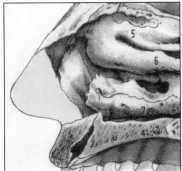

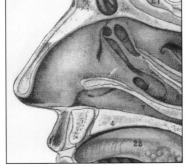

68 | Features and details
Drawing hair

See also
Line and wash, pages 30-31

Drawing hair is fun; it is a tremendously varied subject, and presents all sorts of technical and artistic challenges. The one thing you do not want is to produce a three-dimensional face and then to flatten the whole effect by drawing two-dimensional hair. The correct choice of medium can be crucial in this artistic process.

Hair and ears
Sideburns and longish hair (*top*) throw shadows, and partially obscure the ear. Hair tucked behind the ear (*below*) tends to emphasize it.

Depicting hair
Washes are an effective way to depict hair. The artist has employed a tonal wash combined with line work (*above*), and has used long brush strokes to mimic the flowing hairstyle (*right*).

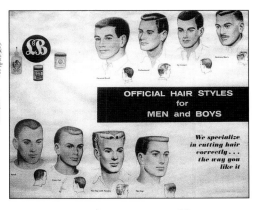

Hairstyles
Hair can be piled up, cut, and arranged in many different ways for visual interest and variety.

Facial hair
The direction of hair growth may distort the appearance of a model's mouth. Draw beards or moustaches with care. Also observe the beard-growth shadow on the male face (*below*). This provides additional colour and texture.

Dark hair
Even in very dark hair, there are cast shadows and highlights that help describe the form.

Using two media
A wash blocks in the mass of head hair. Overlaid pen and ink suggests finer strands.

When using words to describe an individual, we sometimes rely on descriptions of their mood or demeanour: beaming, sad-eyed, down at mouth, wide-eyed amazement, and so on. Similarly, in drawing and painting, an accurate depiction of facial expressions goes a a long way towards conveying the personality of the subject and adding to a recognizable likeness.

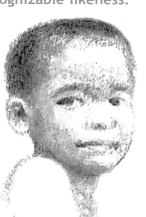

Leonardo da Vinci (1452–1519) **Mona Lisa (La Gioconda)** Italy, 1503 Oil on panel Classic tranquillity, with the "enigmatic" smile.

Tranquil expressions
In contrast to dramatic expressions, a calm, serene facial pose allows you to take more time on your work than the quick demands of extreme emotions.

Smiling
Most people can hold a smile easily, since they are using few facial muscles.

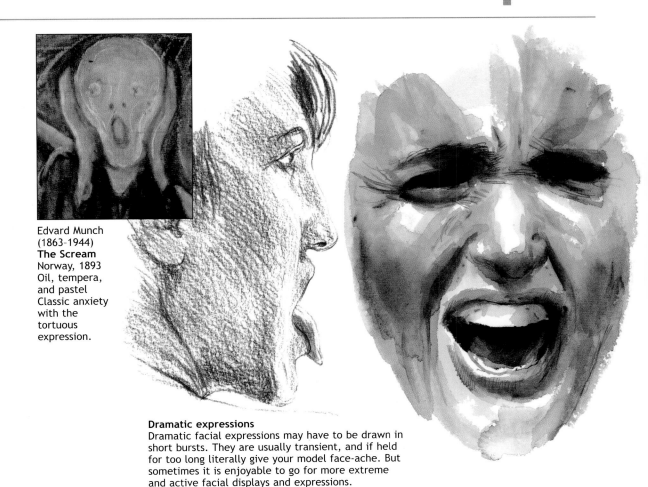

Edvard Munch
(1863-1944)
The Scream
Norway, 1893
Oil, tempera,
and pastel
Classic anxiety
with the
tortuous
expression.

Dramatic expressions
Dramatic facial expressions may have to be drawn in
short bursts. They are usually transient, and if held
for too long literally give your model face-ache. But
sometimes it is enjoyable to go for more extreme
and active facial displays and expressions.

It is possible to recognize emotion in a drawing of eyes, even without including any other features. Astonishment or surprise, for example, is indicated when the eyebrow and eyelid are raised, revealing the white of the eye above the pupil. A startled expression seems exaggerated when the subject is looking down, because the eyes tend to open even wider. Other emotions clearly communicated by the eyes are those of disbelief and, of course, laughter and happiness.

Happiness
Laughing or smiling creates half-closed eyes with raised eyebrows. Look out for laughter lines – the creases on each side of the eye.

Narrow eyes
A narrowing of the eyes suggests the subject is in quizzical mood. Squinting into strong light produces a similar effect. Note also the lowered eyebrows.

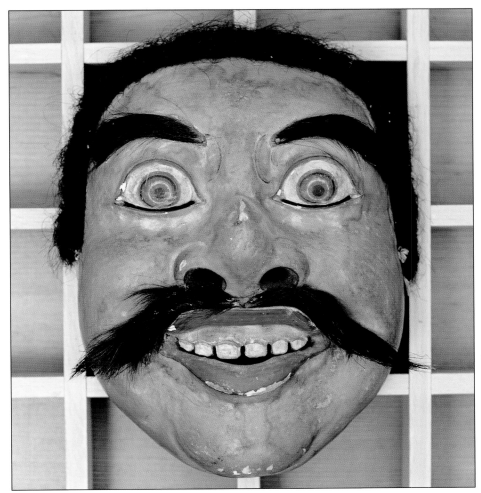

Happy face
Wide eyes, bared teeth, and an upturned smile have been emphasized to convey the mood of happiness in this theatrical mask from Asia (*left*).

Noble Ladies of Mycenae
Greece, c.1300BC. Wall painting

See also
Pure colour, page 31

The best way to approach the task of capturing skin colour is to build up the total picture gradually, by careful observation of the complexion and features of the person. Short cuts, such as filling in with flat colour for skin tones, is fairly ineffective and often looks unrealistic.

Charcoal and pastel
Drawing softened with a paper stump (*below*).

See also
Pure colour, page 31
Eyes: colour and reflection, pages 44-45
Hair, pages 68-69

Expressions and complexions
Colour and tone | 75

The drawings opposite show a variety of techniques:

Watercolour pencil
After the features have been freely drawn with pencil, the skin tone and dark hair are implied with washes *(top left)*.

Graphite pencil
A linear and hatched style of drawing, using a relatively soft pencil *(top centre and bottom left)*.

Pen and ink
A bold and simple sketching style in sepia with hatching *(top right)*.

Brush drawing
Coloured washes are overlaid to create the colour and mass *(centre row, left and middle)*.

Pen and ink with wash
Delicate pen crosshatching with watercolour washes *(centre row, right)*.

Charcoal
Soft charcoal modelled and blended with a paper stump *(bottom row, right)*.

Coloured chalk
The textured paper suggests reflected light in this tonal study *(bottom row, centre)*.

Materials and techniques
Some of the subtleties of black skin created through the use of pastel and stump *(left)*, graphite pencil *(far left)*, oil pastel and wash with scratched highlights *(below)*, and straightforward brush drawing with washes *(below left)*.

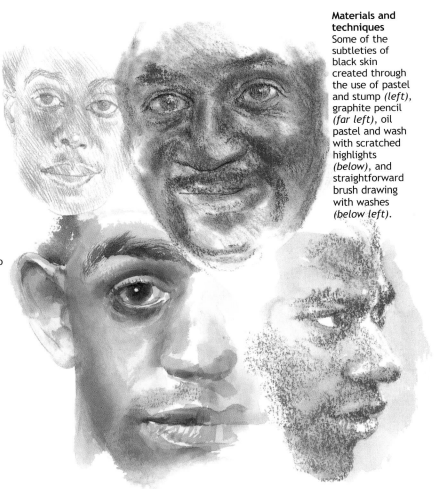

in the portrait studio

Drawn to portraiture

Toby Wiggins is an established and highly regarded portrait artist whose award-winning paintings and drawings have appeared in numerous prestigious exhibitions nationwide.

Wiggins is an artist who acknowledges the time-honoured traditions of figurative painting, and portraiture has become the enduring theme of his work. He has given generously of his time, and opened his studio, to convey many of his methods and techniques, and the creative and technical challenges facing the professional portrait painter of today.

The studio

Toby Wiggins's current studio is in an idyllic and remote setting in the Dorset countryside (in the west of England). The building is in the grounds of a classic English manor house dating back to the 16th century. The double-height, stone-built structure is a former milking parlour that was converted into an artist's studio in the early 20th century, and was formerly occupied by British sculptor Mary Spencer Watson (1913–2006).

The studio has been enjoyed by three generations of artists and is imbued with an atmosphere of creative endeavour. The huge north-facing window provides the quintessential "artist's northern light". Scattered around is a wealth of visual stimuli – books and tools, sculptors' maquettes, plaster-casts, paints and brushes.

The studio is an inspiring and tranquil environment, one that is conducive to creativity and concentration, and all the requirements of a dedicated portrait painter.

"I like working in natural light; this studio represents over 80 years of continuous artistic practice."

Ordered chaos

Amidst this ordered chaos are many portrait paintings in various stages of development; some in progress, and some wrapped, ready for shipping to various clients and exhibitions. Several blank, primed canvases, boards, and gesso panels are waiting for work to start and for images to appear on their surfaces.

The model arrives, and the artist thinks about the pose. He already knows this is going to be a short study, and has already decided to concentrate on the head and shoulders.

He considers the position of the model, the viewpoint and the inclination of the head, the light, and the general relationship between the model and the environment, always taking into account what is going on around and behind the model's head.

Even though the final composition focuses on the head and face, all these elements are important considerations that will contribute to the texture and atmosphere of the work, and the success of the finished painting.

Using a viewfinder

When making a portrait study some people may find it useful to use a viewfinder. This is a classic and basic artists' device – just a simple frame made of cardboard – but it can be very helpful in focusing attention on the subject, especially when considering the compositional possibilities and the crop.

A viewfinder also helps to "read" the play of light on the subject, and to see the broad light and dark areas within the composition. When teaching, the artist encourages students to make preparatory line drawings to understand rhythm, shape, and proportion, and to make monochrome studies to understand what is happening with the light, in the scheme of things; but he says,

"Eventually this all becomes natural and, rather like driving a car, painting ultimately becomes an intuitive process."

82 Support, medium, and palette

For this study the artist is working on a small canvas that is washed with a ground of warm Venetian red.

This ground colour is a traditional starting point for many artists, and on closer study of many Old Master paintings there is evidence of a similar warm red-earth colour visible between the brushstrokes and under the paint surface.

The purpose of the ground colour is to reduce the stark white of the canvas to a mid tone, which enables the artist to work in both light and dark colours. The underlying red also provides a vibrancy and resonance to the subsequent colour application as the painting builds up.

In the subject before him the artist sees a colour range of blue, grey-violet and a little red and violet, and, although by his own admission, it is all "very subjective", he sees an overall blue-violet theme emerging in this portrait study that will act as a complementary colour theme to the Venetian red ground of the canvas.

Palette
For his palette, when working in the studio, the artist uses a melamine table, which is his work station and colour-mixing surface. He always works in oils and has no particular brand allegiance, except to say that his colours must be of "artists' quality" for the obvious reasons of strong pigmentation and longevity.

Brushes

The artist's painting technique allows his works to be touch dry in a couple of days. He employs a range of brushes, from a massive (when seen in relation to the small canvas) 6.5cm (2½in) square brush to a small bristle filbert.

The majority of early work takes place using a large 6.5cm (2½in) ordinary household paintbrush.

The first marks are made broadly with a thin, dark paint application, a colour that Wiggins describes as a desaturated blue-violet. He says the first phase is to consider the shape and tone of the work, and to get the darks and lights in the right place. Broad, dark marks akin to those of an abstract painter appear on the red ground of the canvas, but on closer observation a definite portrait form gradually becomes apparent.

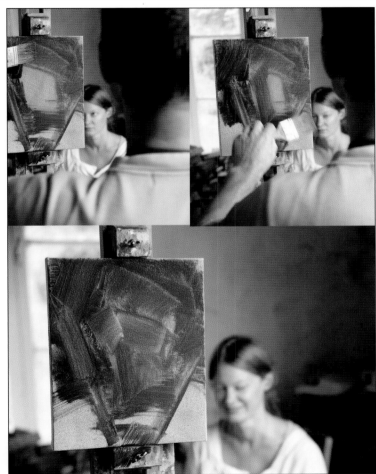

The dark, seemingly abstract brushwork is quickly followed by similar, lighter marks as the artist adds white to his initial dark colour. Although the painting is still rapid and abstracted, the definite form of a head and face is clearly beginning to emerge on the canvas.

"I am trying to define the darkest and lightest areas, while also developing a sense of form and resolving the basic shapes of the composition."

Once the artist has established the foundations of the portrait, he picks up a smaller brush (well, relatively smaller, as it is still a broad 2.5cm (1in) square-tipped brush). This phase he describes as "refining and defining". He explains this as a continuing and simultaneous process of resolving and refining shapes, the composition, and the tone, bringing in colour, and constantly reassessing the balance of the painting, while trying to maintain the freshness of the first marks and the sense of light, likeness, and form.

"The process is not something I consciously think about when I work; I am just doing it."

"The key element is not to be too self-conscious; there has to be pleasure, it has to be an exciting process. Also, you have to be quite robust; you need to take hard decisions and be prepared to make sacrifices. If marks are wrong they have to go, even if you like them. You have to be rigorous and self-critical; it is quite a tiring process."

"I can paint for five or six hours and make good progress, but I eventually reach a point where I have to put the brush down."

The model needs a break too
The artist marks the position of the seat on the floor with tape before breaking the pose.

Work in progress
The artist and the model review the work so far.

"That mark jumps out! This is all the wrong angle!"

During model breaks Wiggins casts a critical eye over work in progress. He takes the work off the easel and holds it upside down, viewing the inverted image in the large studio mirror.

"Standing behind the easel I become familiar with every mark, and now looking at the work upside down and back to front enables me to distance myself and see the painting in an entirely different way."

In the portrait studio
Thinking about colour

"I am looking at the colour of the subject, the colour on the palette, and on the canvas. I can pick a colour off the palette and put it on the canvas, and find it influences and changes the appearance of everything in the painting."

"There is a shadow here; the eyelid is very fine at this angle and the light is just catching it. I am being quite particular at this stage."

The painting process, says Wiggins, is quite intuitive, and we arrive at a stage that the artist describes as being far more rigorous than the earlier stages as he defines the character of the sitter on the canvas.

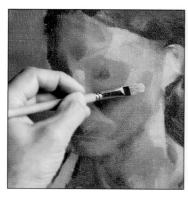

Using increasingly finer brushes, the artist resolves the detail of the mouth and the light in the eyes.

After working continuously for three hours, he says he has reached a satisfactory point where he has achieved various fundamentals in the painting, but can push the portrait further. He feels there is a likeness emerging, and that he is getting a sense of the light.

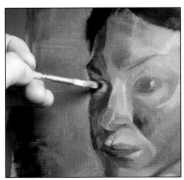

See also
The finished portrait,
pages 102-103

"I will give it another half an hour. I have resolved the basic form, but there are many adjustments to make."

"The neck is looking a bit blocked out; the mouth is twisted and not formed properly. I will put more light on the cheek and more light in the eyes to convey the sense that the eye is glassy – the model's eyes are a lovely blue-grey and quite luminous. I'll work on the mouth and try and get that correct, and I'll work on the whites and highlights, and make a few adjustments to those areas."

"I've reached a point where I would probably stop for the day. I'll leave it open for a day or so, consider it, and then push it further."

Toby Wiggins
Study of Catherine No. 4, 2007
(detail, work in progress)
Oil on canvas
30 x 24cm (12 x 9½in)

Self-criticism
"A reasonably successful session for three and a half hours. I like the quality of the paint marks in certain areas, but there is not enough light and luminosity on the canvas yet. It is not quite there; the form of the face is not exactly right, the peculiar angles are a tiny bit wrong. Subtle changes need to be made. The lips are irritating me now, and the shape of the nose is not quite there (*at this point Wiggins adjusts the wet painted nose with his finger*); I'll leave it for a day, think about it, and see what comes up to the surface. I sometimes give a painting a light sanding and work it over again, but all in all I am quite pleased with this little oil sketch. On the other hand, though, another three hours would not hurt."

98 Thinking about likeness

"Technically I approach the problem of likeness through establishing a sound structure. This means thinking a lot about the skull."

"Getting the form and inclination of the skull correct from the outset can take you a long way towards obtaining a good likeness."

"The skull (cranium) is fixed, but it rotates on the neck and strikes an axis from the base. This inclination or tilt of the head is unique to each sitter. The human skull only has one moving part – the mandible (jawbone) – and over this bone structure are stretched all the muscles.

The facial expressions are muscular, and to establish the correct proportion, position, and shape of the muscular features over the skull takes you further still in a quest for the likeness and character of an individual.

I am not a painter who is intent on flattery or on caricature. I hope that I do not try to make a person look a certain way; they are what they are.

Within each individual's face, I find certain shapes and rhythms emerge. The curve of an ear is found again in the line of a nose or shape of a nostril. Forms repeat themselves.

There is always some detectable asymmetry within the human face – the slight squint of an eye, or the subtle twist of a nose or chin. Noticing these things helps to develop 'likeness'.

Skin, hair, and eyes are very different from one another in terms of tone, texture, and colour. I try to render the differing ways that light reflects on the surface of these elements.

I think that through this business of looking intensely at what is there in front of you, the character of the sitter emerges."

"When someone is sitting, they remain in a state of animation; there is always some movement within the face. This movement and energy must always be taken into account."

"I see the qualities of the sitter emerge as my work progresses, and I try to make a good, honest portrait."

Toby Wiggins
Study of Catherine No. 4, 2008
Oil on canvas
30 x 24cm (12 x 9½in)

"I tried to keep a delicate balance between enriching this little painting and not working it to death, which is always a very real danger."

After the initial sitting of three hours, which is documented on pages 80–99, Toby Wiggins put aside the study of Catherine and left it to dry for some weeks.

In the earlier sittings Wiggins successfully established a good sense of form, light, and tone, and built the solid foundation of the portrait. The artist returned to the painting when he and his model could find a few spare hours for more sittings (note how the model's neckline has changed), and two more sessions of about three hours each took place over the following few weeks.

During these two later sittings Wiggins says, "I attempted to put more 'flesh on the bones' as it were, and to develop a deeper sense of the model's likeness."

It is at this stage that he allowed his whim to dictate which aspects of the portrait would become important: "In this case I particularly concentrated on capturing and resolving the subdued light and the luminosity of Catherine's skin and eyes."

Additional sittings allowed for further development of all aspects of the portrait, but at this stage the artist is always aware of trying to maintain a delicate balance between enriching the work and not overworking it, as this is always a danger.

"My intention is to keep the brushwork fluid and alive. In this study there is a predominance of greys and desaturated colour notes; I am always working to keep these colours clear and fresh to avoid the portrait from becoming too overdone."

The whole oil portrait study from start to finish took about ten hours to complete, and Wiggins says, "I feel quite pleased with the result, and even after several months I can return to this little canvas and not feel disappointed with it. This is a good sign."

Photographic reference conveys pose and crop possibilities, and provides more detail of the colouring and complexion of the sitter. (The artist did not use photographs when painting the three-hour studio colour sketch on pages 80-102.)

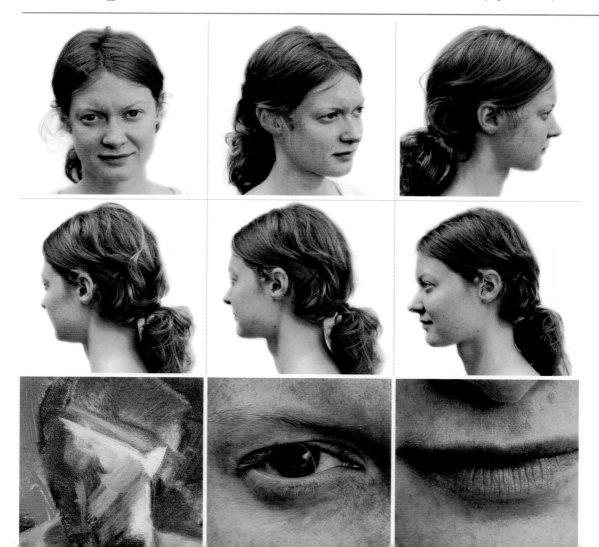

See also
Recording features and details, pages 62–63
Visual reference, page 178
Photographic sources, page 179

In the portrait studio
Model reference | 105

106
In the portrait studio
Influences and inspiration

"The Master paintings are here for everyone to see, and as a painter I am always learning from them. I do not look at them solely for painting technique; it is the level of humanity and vision in them that I find so immensely inspiring."

"I come back to certain paintings in the National Portrait Gallery and the National Gallery in London time and again. These two major galleries have among the greatest collection of paintings in the world; you cannot be disappointed visiting them. The pictures there are phenomenal. Portraits that are hundreds of years old can seem so fresh and have such 'inner life' that the person depicted almost breathes.

I like to examine the Northern European portraits, including such works as van Eyck's *Arnolfini Portrait*, and those by Spanish painters such as Jusepe de Ribera, and Velázquez.

Another particular favourite of mine is Titian's *Bacchus and Ariadne*. While not strictly portraiture, this painting displays the incredible skill of a virtuoso

painter. The dynamic composition, with figures caught in flight, has such rhythm and intensity of colour. I have looked at this painting since I was a teenager and I still see it with fresh eyes today. The level of ambition in this work is remarkable."

Titian (*c*.1487-1576) (*right*)
Bacchus and Ariadne, 1520-22
Oil on canvas
(whole canvas, unframed)
National Gallery, London

Titian (Tiziano Vecelli or Vecellio) was the leading Venetian painter of the 16th-century Italian Renaissance. His versatility, and wide range of work – in portraiture, landscapes, religious paintings, and mythological subjects – brought him international fame.

Jan van Eyck (died 1441) (*left*)
The Arnolfini Portrait, 1434
Oil on canvas
National Gallery, London

Jan van Eyck (or Johannes de Eyck) was an Early Netherlandish painter from the Maastricht area. He is credited with introducing modern techniques of oil painting, which enabled the artist to depict subtle effects of light.

thinking about
self
portraits

Desmond Haughton
Self-portrait with Yellow Waistcoat
(1994)
Oil on canvas
86.5 x 51cm (34 x 20in)

Desmond Haughton is an artist who
has painted several self-portraits
from a simple colour palette. Here
he uses Lemon Yellow with Titanium
White and Burnt Sienna for tints
and shades to create the striking
yellow waistcoat around which the
composition revolves. The choice of
colour and style of costume
influences the design and mood of a
self-portrait painting and conveys a
tremendous amount about the
character of the sitter.

*"If you are not honest to the
colours you see and the shapes you
see, a painting can start to go
wrong. It is all about logic, a
logical sense of seeing."*

(Previous page)
Self-portrait oil sketch
(2002)
*The artist at work in
his London studio on a small
self-portrait oil sketch.*

Artists have always made self-portraits. In the 6th century BC the architect Theodorus of Samos cast a bronze portrait of himself, holding in his right hand a file, and in his left a model of a chariot he had made. Medieval artist scribes used to draw themselves in the margins of their manuscripts, often with some complaint about the difficulties of their task.

Among the great self-portraitists was Rembrandt, who in 60 paintings and some 20 engravings converted the self-portrait from an artistic exercise into a visual instrument of psychological interrogation. His self-portraits reveal for all posterity every phase of the artist's life, and this body of work must certainly have influenced fellow Dutchman Vincent van Gogh (see the following pages), some 200 years later.

Rembrandt van Rijn (1606-69)
Self-portrait aged 34
(1640)
Oil on canvas
103 x 81cm (40³/₄ x 32in)
National Gallery, London

Rembrandt is generally regarded as the greatest Dutch painter and one of the masters of European art. His lifelong series of self-portraits reveal his changing circumstances from eager youth to later hardship.

Lorenzo Ghiberti (1378-1455)
An Italian artist of the early Renaissance best known for his sculpture and metalworking, Ghiberti was born in Florence. His great masterpieces are the two sets of bronze doors made for the cathedral Baptistry in Florence. The second set, in more naturalistic style, had a profound influence on the artists and sculptors of the Renaissance, many of whom trained in his workshop. Ghiberti also incorporated his self-portrait (*above*) into the design of the doors.

This is the most recently discovered portrait photograph of the young Vincent van Gogh. Research suggests that it was taken when Vincent would have been between 16 and 20 years of age.
(*Courtesy of Richard van Dijk*)

"These photographic portraits wither much sooner than we ourselves do, whereas the painted portrait is a thing which is felt, done with love or respect for the human being that is portrayed."
Van Gogh to his sister Willemina

In the course of four years, between 1885 and 1889, Vincent van Gogh (1853–90) painted more than 40 self-portraits. This is a unique and an extraordinary achievement, not just in terms of artistic expression and experiment, but as an intimate record of the pilgrimage through life of a human being.

Van Gogh's intense visual self-analysis was spread over a short time – four years – and on the whole this work is deeply interrogative. He had always been suspicious of photography, even as an instrument of record. Only three portrait photographs of him exist, all taken before the age of 21, and he does not face the camera in the few other photographs known of him.

His first moves into portraiture were mostly motivated by economic incentives, and towards the end of 1884 he felt that he could make a living by painting portraits. It was at this time that Vincent started painting self-portraits. His reasons for doing this were mixed. Foremost, there was the question of expense: models needed to be paid in one way or another, and it cost nothing to paint himself.

It would, however, be a total misapprehension to see in van Gogh's pictures of himself nothing more than exercises in studied self-presentation. Few men have shown so painstakingly, in their writings and in their paintings, the search for the nature of their identity. The very number and diversity of his self-portraits is proof of this endeavour.

Vincent van Gogh, Self-portrait
inscribed *"à mon ami Paul G"*.
Arles, September 1888
Oil on canvas
61 x 50cm (24 x 19½in)
Fogg Art Museum, Cambridge, Massachusetts

Van Gogh painted more than 40
self-portraits over a four-year
period. On the following pages are
details of 18 paintings chosen for
viewpoint and costume.

Self-portrait as an artist
Paris, early 1888
Oil on canvas
65.5 x 50.5cm (25¾ x 20in)
Vincent van Gogh Foundation
Van Gogh Museum, Amsterdam

It is a sad and ironic fact that the last self-portrait that van Gogh painted was a lie. Early in 1889 his mother celebrated her seventieth birthday. Although they had not been in constant contact, Vincent still cherished the warmest memories of her, and they kept up an irregular correspondence.

Later in the year he decided to send her this self-portrait, mainly to reassure her about his health. Van Gogh painted himself not as he then was, but as a much younger and healthier-looking man: clean-shaven, with well-brushed hair and a proper painter's smock. He was, however, unable to avoid the look of desperation that hovers over his features.

Self-portrait
Saint-Rémy, September 1889
Oil on canvas
40 x 31cm (15³/₄in x 12¹/₄in)
Private collection, Switzerland

These lightly sketched, but sharply observed, drawings from life became the basis of a whole series of subsequent oil paintings. These drawings are the only surviving examples of preliminary work on Vincent's own portraits, which he usually approached directly on the canvas. They do, however, give the lie to those critics who accuse van Gogh of technical incompetence.

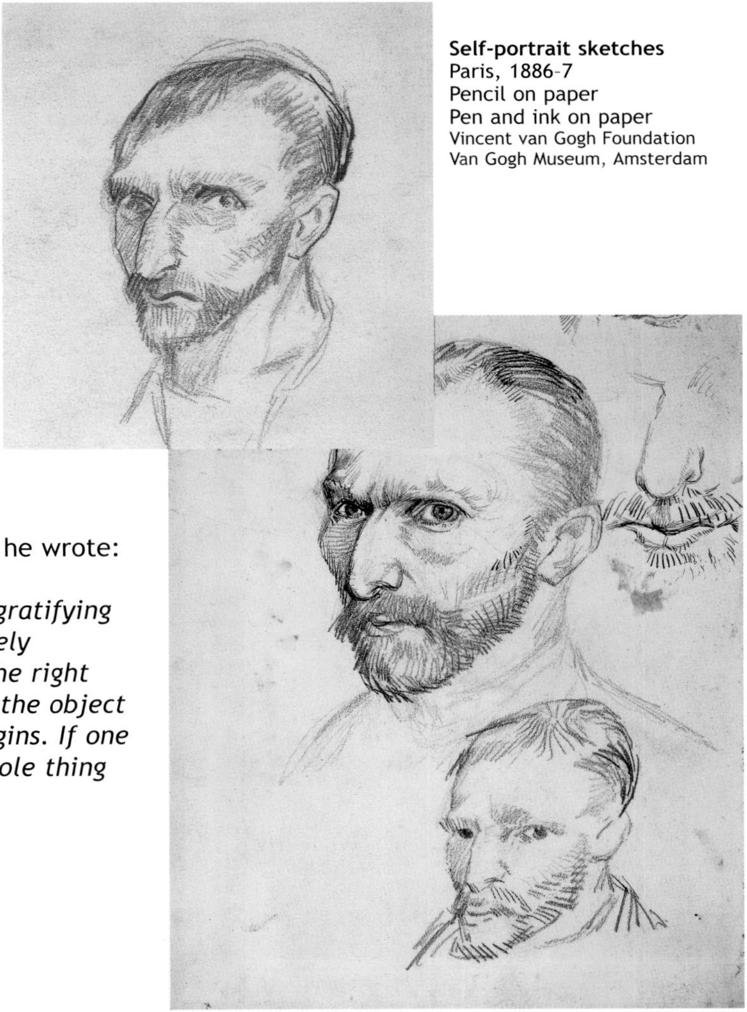

Self-portrait sketches
Paris, 1886-7
Pencil on paper
Pen and ink on paper
Vincent van Gogh Foundation
Van Gogh Museum, Amsterdam

In a letter to his brother Theo, he wrote:

"In a word, [painting] is more gratifying than drawing. But it is absolutely necessary to be able to draw the right proportion and the position of the object pretty correctly before one begins. If one makes mistakes in this, the whole thing comes to nothing… "

Self-portrait, 1906
John Singer Sargent (1856-1925)
Oil on canvas
71 x 53cm (28 x 21in)
Uffizi Gallery, Florence

To make a finished, concentrated portrait study, you will need the cooperation of a patient model; failing this, draw a self-portrait. You are your own best model, willing to sit at a moment's notice for free, and – hopefully – not upset anyone other than yourself if you produce a portrait that is inept or unflattering. The famous society portraitist John Singer Sargent once observed,

"Every time I paint a portrait, I lose a friend."

As an art form, self-portraiture has immense appeal. It is an opportunity for self-analysis; a chance to study a reflection that one sees fleetingly in a mirror. The penetrating gaze that is common to nearly all self-portraits no doubt reflects the intense concentration of the artists wrestling with the problem of capturing an image while looking deeply into their own eyes.

Setting up a self-portrait
Many self-portraits are full length and show the artist at work with materials in hand or at an easel. You will need a reasonably large mirror for full-length self-portraits; free-standing, pivoted mirrors, such as those used in dressing rooms, can be arranged to capture different angles and to create a sense of depth and space, and to make a frame within a frame.

☛ **Self-studies with body**
In these self-portraits the artist (see also the following pages) confronts the mirror full on at three-quarters length. He used a variety of media to capture these quick studies.

"It is not so much how others see me, but more, how I see myself."

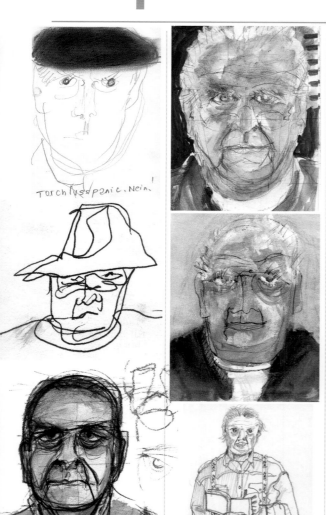

Torchlyssponic. Nein!

Are you there. I want you to meet I said meet me down by the Belong

SELF-PORTRAIT 9-II-06
WINE & WORCESTER SAUCE

YOU ARE WHAT YOU EAT

Self-studies in sketchbooks may be made with a variety of media (usually whatever is available at the time), including ball-points, fibre tips, pencil with watercolour tints, correction fluid, and even red wine. Here the artist has used a touch of Worcester sauce to tint a sketch (*left*).

Experimentation
A big advantage of self-portraiture is that you have plenty of time to experiment with viewpoints, lighting, and different media. Here and on the previous pages the artist has drawn himself using lots of different materials. These sketches span a number of years and capture many moods and situations; they act as a private visual diary and aide-memoire.

art
directions

As an artist who works directly from life on copper plates, Gemma Anderson says: "My portraiture is a very particular art and one that needs a constant supply of models." To nourish her quest to represent the truth fairly about someone, a supply of willing people is essential and she has turned to friends and family as the subjects of her huge-scale, dramatic suite of etchings entitled *People I Know*.

Gareth

When Gemma began these portraits she says her starting points were *"the people I know and the things I imagined inside them"*.

Anderson's portraits are meticulously careful and she is very skilful in gaining a likeness of the person in front of her, but on closer inspection the viewer will often see curious things going on, things that are not initially apparent.

At first glance what may appear to be a double portrait is really the sitter with their secondary or alternative personality, which Anderson sees, physically emerging alongside. Vegetation grows out of people's heads, objects are held in hands, emblems and symbols appear on clothing. Normally invisible veins and organs become discernible. The artist says: "I started making these life-size etchings of people I know, but gradually introduced forms

that both physically resemble their anatomy and metaphorically represent something about their character, their personality and their idiosyncratic self."

James Unsworth

Etching people
"People have mostly been documented in art through painting and photography. The natural history and medical context of etching interests me as its strong descriptive line is perfect for comparing the anatomies of people, plants, and animals, and these etchings also become emotional maps of the sitter."

See also
Working with a model, pages 22-23
Glossary of terms, Etching, page 183

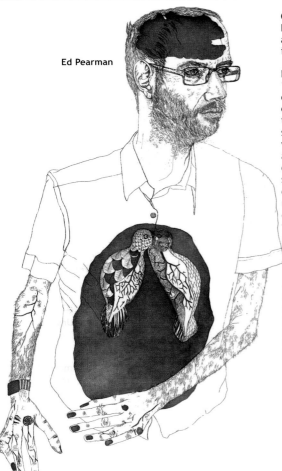

Ed Pearman

Gemma Anderson
Hand-coloured etchings, 2007,
all 100 x 80cm (39$\frac{1}{2}$ x 31$\frac{1}{2}$in)
in editions of 10

No mistakes
"Working directly onto the copper plate from life was challenging as I would carry the plate to meet the sitter, the plate being very big and heavy, and prop the plate up on a chair ad-hoc style. Drawing onto copper allows for no mistakes; each line I draw is etched, so there is an excitement to the process which intensifies the line. Once I have captured the form of the person I then over-draw any idiosyncrasies of anatomical details, symbols, and objects particular to the sitter, and then I etch the plate and print it."

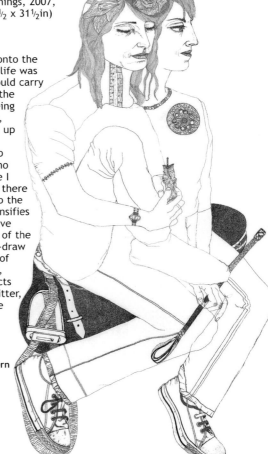

Laura Hern

128 About face

George Underwood
Shades of Green, 2007 (*right*)
Oil on canvas, 91.5 x 76cm (36 x 30in)
The men (*left*)
Oil on canvas, 91.5 x 63.5cm (36 x 25in)
Eighteen paintings (*following pages*)
Oil on canvas (*details*) (various sizes)

George Underwood is an artist who has always been concerned with portraying the human face, and faces have become a major recurring subject in his work. He has developed this in a much more imaginary way than in straightforward observational portraiture.

George says, "One of my current themes is – for want of a better description –

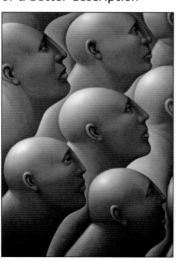

collections of heads and faces. These people mostly seem to come from different eras and backgrounds. I have no conscious idea who they are, where they come from, or to what age they may belong." George recalls that as far back as he can remember he always wanted to draw or pick up a paint brush, and when he was very young he enjoyed drawing faces with his finger on the steamed-up kitchen window.

One of the artist's first art heroes was Salvador Dali, and when he was about 13 he started to consider art as a serious career option. At the age of 16 he went to art school. "Back then," he says, "the general belief was that there was no money to be made in just painting." So he was channelled into "commercial art", pursuing an early career as an illustrator during which he

produced literally thousands of book covers, LP and CD sleeves, adverts, portraits, sculptures, and drawings. This gave him the valuable opportunity to experiment with many different subjects and materials, although he says, "painting was what I really wanted to do all along – just draw and paint and be good at it".

George started painting in oils in the early 1970s and now often exhibits his work in impressive one-man shows and group exhibitions, such as The Royal Academy in London. He is fascinated by the way certain artists have mixed fantasy and realism in their work and acknowledges the influence on him of the 20th-century artists of the Vienna School of Fantastic Realism, as well as late-medieval Flemish visionary painters such as Brueghel and Hieronymous Bosch.

Graffiti, as we see it today, was born in the New York subway in the 1970s. Some regard it as "high art", others as senseless urban blight; but, whatever your opinion, anonymous "artists" have been drawing on walls for thousands of years. The human face is an enduring subject for the "graffitiist" who, with the help of the aerosol spray can, and with whatever other tools come to hand, has made the face a familiar image that defaces walls all over the world.

Graffiti or mural?
Attributed to the artist known as "Banksy", the narrative of the wall image (*left*) needs little explanation, and the expressions on the protagonists' faces (*below*) are very carefully observed. The artist's work has become highly valued, and blurs the distinction between subversive street art and establishment gallery art.

The agony and the ecstasy
Some consider graffiti as art, but others, particularly those who have to clean it off buildings, see it as an act of irresponsible vandalism.

Art or anarchy?
The following pages show a selection of street art seen in Granada, Spain.

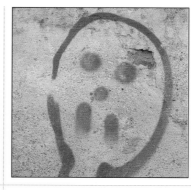

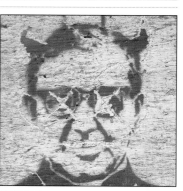

136 | Inspired by graffiti

The artist explores image ideas inspired by the New York graffiti artists of the 1970s and 80s. Here text, or "tagging", is combined with loose and spontaneous brushwork in paintings that pay homage to the American artist Jean-Michel Basquiat (1960-88).

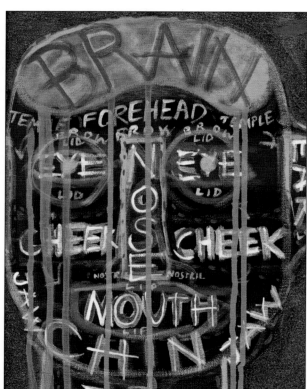

Talking heads
(*right*)
Acrylic on
canvas
131 x 104cm
(51 x 41in)

Doctor, doctor
(*opposite page*)
Acrylic and mica
on canvas
104 x 74cm
(41 x 29in)

Sketches and
development
drawings for the
paintings were
done in charcoal
on paper.

☛ **Semiotics,** the study of signs and symbols and their use and interpretation.

☛ **Pictograph** (also known as a pictogram), a pictorial symbol for a word or phrase.

A pictograph, also known as a pictogram, is a pictorial symbol used for a word or phrase. Pictographs were the earliest known form of writing, and examples dating from before 300BC have been discovered in Egypt and Mesopotamia (present-day Iraq).

High noise levels
Use ear protectors

In the latter part of the 20th century, particularly from the late 1950s onwards, graphic designers revived an interest in the use of pictographs as a practical means of communicating and to reinforce the verbal or written message.

Massive post-war reconstruction, with new public buildings, roads, and mass transport systems, combined with the continual movement of people across frontiers, demanded simple and effective means of non-verbal communication. Thus many artists and designers became involved in utilizing *semiotics,* which is the study of signs and symbols and their use and interpretation.

Pictographs were demonstrated to be particularly effective in grabbing attention and in communicating information simply and directly.

Among the thousands of instructions, directions, and identification messages that architects, planners, and designers of the built environment had to consider was the problem of simply portraying the human form.

We are all familiar with the male and female pictogram identifying toilet facilities, and the walking figure who is seen at pedestrian crossings worldwide. Designers also had to consider the stylization and simplification of the human face and head. Some of the varied and creative attempts that have been made to achieve this aim are shown here and on the following pages.

Language barriers . You do not have to be fluent in either English or Spanish to understand what the pictographs in these photographs (*left and right*) are trying to say.

Pictographs
A safe and effective means of communicating the head and face, pictographs are a major aid in overcoming language and literacy problems.

Art directions
Graphic faces | 139

Sign survey (1960)
Henry Dreyfuss

Henry Dreyfuss (1904-72), an American industrial designer, was a pioneer of rational, harmonized design systems as a practical aid to communication and greater international understanding. He conducted a huge visual survey of pictographic signs in the 1950s in his work to create a non-verbal graphic language.

From left to right

X-Ray department
Hazardous/Danger
ENT department

Shower room
Drinking fountain
(two versions)

Music room
Theatre
Press interview room

Television room
Group meeting room
Hairdresser (female)

Sign language
Examples of the human head and face
from Dreyfuss's archive of over 20,000
signs and symbols collected from
around the world.

Art directions
Graphic faces | 141

From left to right

Head nurse
Laryngology dept.
Neurosurgery
Speech and hearing

Phonetics/speech
Head of department
Neurosurgical dept.
(Encephalography)
General surgery

Hard-hat area
Wear ear protection
Wear eye protection
Wear respirator

Baggage handling area
Customs inspection
Food preparation
Physiotherapy dept.

Modernist eyes
Museum studies based on works of
some masters of 20th-century art
Charcoal in sketchbook
Original page size:
21 x 30cm (8½ x 11¾in)

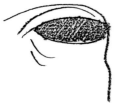

modigliani

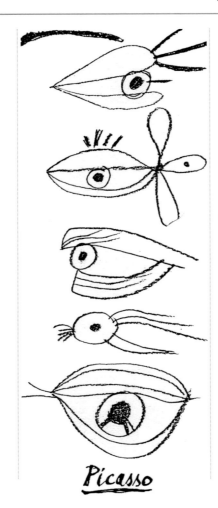

Picasso

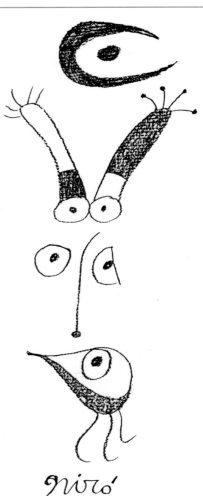

miró

Abstracted eye
(after Picasso)
Coloured glazes on ceramic tile
15.3 x 15.3cm (6 x 6in)

Eyes to the future
Abstracted portraits influenced by the
'Modernist art' movements of
the 20th century

Different perspectives

From the moment of the invention of the photographic negative in the early 19th century, and the ability to fix a photographic image more or less permanently on a surface, photography as we know it today was born. Many contemporary observers commented that "from this moment on painting is dead"; and photography undoubtedly challenged the academic notion of realistic figurative painting, and allowed artists to look at the world from different perspectives. This approach in turn influenced the great modern-art movements of the 20th century, leading to Cubism, Futurism, and Surrealism, of which artists such as Mondrian, Picasso, and Miró were exponents.

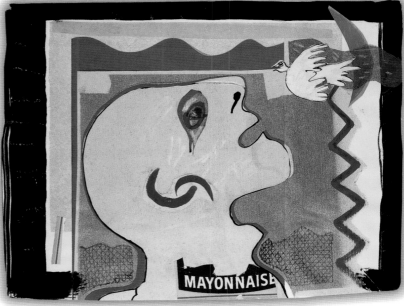

MAYONNAISE

144 Drawing with your eyes closed

EYES CLOSED II (
10·2·

Art teachers employ many techniques to help their students "loosen up" and to approach their work in a more spontaneous, instinctive and unselfconscious way.

One of these techniques is known as "blind drawing", where the artwork is made either by attempting to make a drawing while keeping one's eyes closed throughout the whole drawing process, or alternatively by keeping one's eyes fixed on the model and never looking at the paper on which you are making the drawing.

Both techniques can lead to some fascinating and uninhibited results as the artist has little control over composition and the placing of the elements on the paper, and is therefore unable to correct or alter the unpremeditated marks.

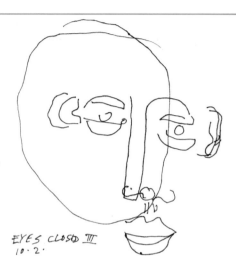

EYES CLOSED III
10·2·

Visual memory

Blind drawing techniques have the added advantage of reinforcing visual memory, and the more you do them the better the drawings can become. The fragmented results that come about through chance and accident are often surprisingly interesting and redolent of the work of the 20th-century Cubist and Surrealist artists who challenged conventional notions of composition and photographic reality.

Eyes-closed painting
Acrylic on canvas
70 x 49cm (27½ x 19¼in)

The artist has developed an eyes-closed study into a painting. The original drawing has been transferred onto canvas and placed in a surreal background, which reflects the strange and accidental rendition of the face. The ornate gold frame is a device usually associated with more conventional, classical portraiture, but here acts as an appropriate organic container for this unusual free-form portrait.

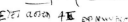

Unpremeditated marks

There are a few techniques that can be applied when making blind drawings. One is to use the other, non-drawing, hand as a fence to prevent the drawing instrument from constantly running over the edge of the paper. Another is to consciously start the drawing from a fixed point – the eyes are a good starting point – and then try to work outwards from there. The finished drawing can be remarkably free, and astonishingly successful.

The art of makeup is an integral aspect of film and theatre, and an artform that is seldom celebrated.

During the heyday of the Hollywood studio system William Tuttle (1912-2007) was head of the makeup department at Metro Goldwyn Mayer, during which time he contributed to a number of Hollywood film classics and had to make people look beautiful for "beautiful pictures".

Colour movies became more dominant from the 1950s onwards, and Tuttle and his team of artists had to face new and unforgiving demands of lighting and colour when working in the latest film medium known as "Technicolor".

New widescreen technologies also had to be taken into account, and the makeup artist had to remember that the face of the performer when seen on screen, in closeup, could be magnified up to 1000 times. Consequently, in the days before imperfections could be corrected by computer manipulation, makeup had to "give" screen actors perfect teeth and flawless complexions.

In addition to creating beautiful faces, dramatic and sometimes horrific characters had to be created, with special facial effects for vampires and monsters, and for the injured and disfigured. Tuttle created the monstrous Morlocks for the film of H.G. Wells's *The Time Machine* (1960), and the stitched-cranium, bolt-necked monster for Mel Brooks's *Young Frankenstein* (1974).

Actors also had to be transformed into a likeness of the characters they were playing. A classic example of

Vincent van Gogh, self-portrait 1887

this facial transformation is that of Kirk Douglas's uncanny likeness to Vincent Van Gogh in the film *Lust for Life* (1956).

The makeup artist's palette
The makeup artist's palette consists largely of an assembly of cosmetics, wigs, greasepaint, and dyes, and the all-important mirror. Over the years Tuttle created plaster masks of the faces of a great many Hollywood stars, which allowed scars, wrinkles, teeth, and false noses to be fitted even when the actors were not present. The masks were part of a system he developed to speed the process of applying makeup during the shooting of a film. About 100 of these masks survive and are now at the University of Southern California.

William Tuttle was head of the makeup department at MGM for 35 years, and is seen here with the University of Southern California's collection of his masks.

Art directions
Film and theatre | 149

William Tuttle 1970, courtesy of the University of Southern California University Archives

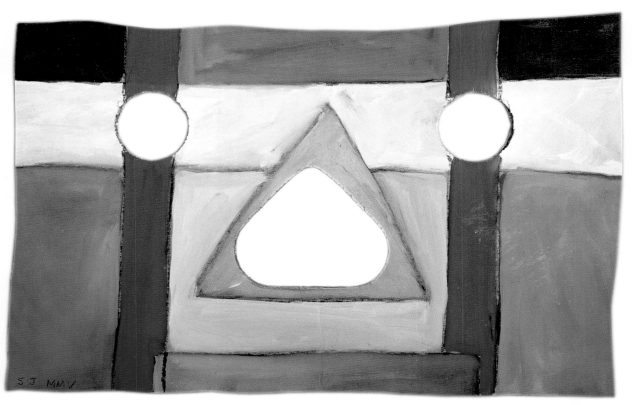

Phizog, number 1, 2005
Charcoal and oil on primed carton
35 x 23cm (14 x 9 in)

Sketch designs
The sketches (*right*) are based on the African tribal tradition of mask design where features are stylized and distorted.

Outsider art
The artist has been influenced by "outsider art", a form of primitive art that is the product of the untrained hand. It is an artform that employs found materials, such as household paints, scraps, cartons, and castoffs. The resulting work can be striking in its rawness and naive originality.

Phizogs
The images here are painted on the end of cardboard banana boxes. Fruit cartons are perforated to provide ventilation and lifting points, and it was these openings, mainly circles and triangles, that drew the artist's attention and suggested a face.

Primitive faces

Sadly, "primitive art" is a negative term; it implies a lack of refinement and of technological facility. However, close study of the sophisticated art and artefacts from low-technology societies, particularly the tradition of mask making, clearly demonstrates that this is not the case. "Outsider art" carries forward the practice of employing low-technology materials, and shown here are three-dimensional examples constructed in the tribal tradition from found materials, thereby creating something from nothing.

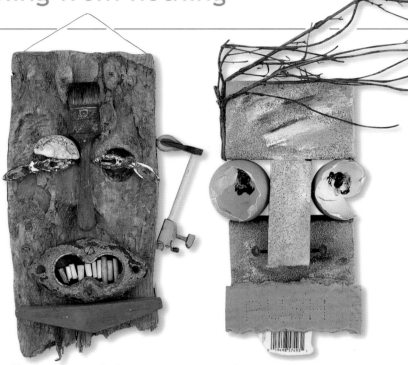

Weather-beaten plank
This head is a piece of stained and weather-beaten plank, with a painted-wood remnant for a chin, the nose an abandoned paintbrush, and a broken piano hammer for an ear. A natural knot-hole forms an eye socket, with old keyhole escutcheons for the eyes. The mouth is a bit of old iron with Victorian claypipe stems for teeth.

Hair of twigs
Polystyrene-foam packing materials form the skull and nose, with hair of twigs, and eyes made of eggshells daubed with paint. The nostrils are old rivets, and the mouth is constructed of cardboard and paper packaging remnants.

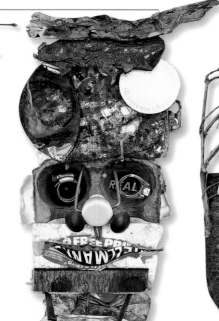

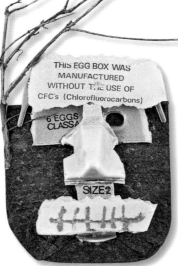

Skull of polystyrene
The skull is made from a block of expanded polystyrene, with a forehead and chin of crushed cans. The eye sockets are egg boxes, with a ring pull and an automotive hose clip for eyes. The nose is a bottle top, with nostrils of rivets, and the mouth is collaged from jar labels. Hair or head decorations are indicated by dried bark.

Road-flattened face
The construction of this face makes it look almost two-dimensional, using a rusty, road-flattened can for the face, with features made from plastic packaging. The lettering on the packaging has been left facing outwards to indicate hair and texture for added visual interest.

Meat-packing features
This face is a polystyrene meat tray, with cheeks and nose made from cardboard corner reinforcements. Beard stubble and eyebrows are rusty nails. The ears are coat-hangers, and wire from champagne corks form the eyes. The whole is painted with acrylic paint.

Photographic and drawn illustrations showing hairstyles from a professional hairdressing manual, *c*.1930.

These illustrations from a 1930s hairdressing catalogue cross the boundaries between handcrafted artwork-illustration and mechanical photography. Printing techniques in the 1930s were not as sophisticated as those of today and it was common-place for artists to skilfully retouch, colour, and overwork photographs by hand to ensure a greater clarity of image to meet the demands of the printing presses of the day.

Nothing new
Ovid (43BC-AD17) was a Roman poet who makes mention of the proud ladies who frequently colour their hair by means of dyes and potions. The Romans well understood the art of hair tinting, and used to render the hair blond by means of chemical lightening agents.

Art of the hairdresser
On the following pages the artist has used line-drawing techniques to convey a variety of hairstyles for the customer to choose.☞

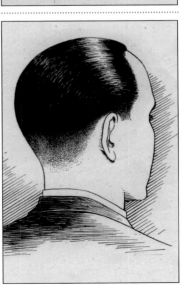

"I am not really a computer person.

I am a typewriter and photocopier person."

Nadine Faye James is an artist who has discovered the joys and eccentricities of an original metal-type printing press. This finding inspired her current enthusiasm for traditional letterpress printing, and to appreciate the physicality of the type forms.

She says she was not particularly passionate about typography in the traditional sense – except to say she was curious about the playful typography of concrete poets and the work of experimental Constructivist typographers such as El Lizzitsky – it was the shapes of the individual characters that really attracted her and she began exploiting these letterforms in her artwork.

She started by making art that integrated large type forms with areas of tone and texture made of smaller "typewriter" areas, and it was about this time that she began making playful little portraits of her friends using only the typewriter. These portrait drawings have became very popular.

The artist says there is a formula for obtaining a good likeness: "I have to look very closely at the face of the sitter, particularly at the nose and the hair.

Some people require a slash and an underscore for their type of nose, others need a capital 'L' or 'J', and with others I can just use two commas for nostrils.

I have learned to pose the model in a position that suits the sitter's facial type and the typewriter's typeface. Hair can be quite a problem as there are no vertical lines in typewriter fonts, but some people are perfect for the typewriter treatment, particularly those with curly hair and big eyes.

It is important to get the space between the sitter's hairline and eyebrows right. Children, with their bigger eyes, require minimal treatment, but older sitters need more character and lines to the eyes and nose; I use a diagonal slash for these. Some older sitters occasionally get quite upset seeing wrinkles."

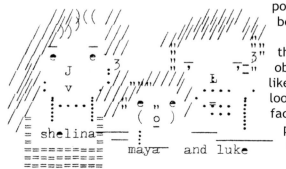

Nadine Faye James
Triple portrait, 2007
Manual typewriter
on cash-register roll
Enlargement approximately 235 per cent
Inset (*opposite page*) is original size

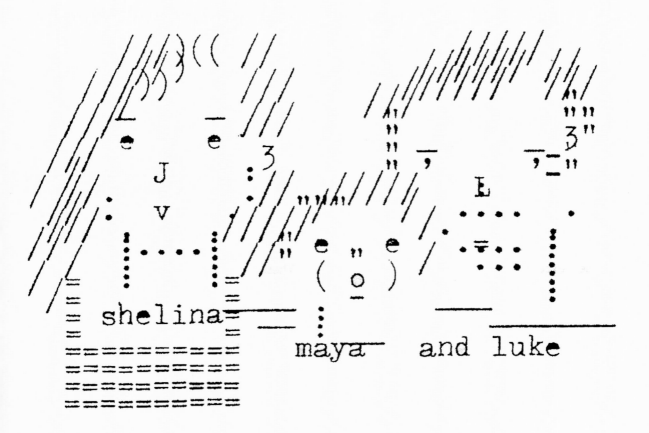

Nadine Faye James
Typewriter portraits, 2007
Manual typewriter
on cash-register roll
(*Exact original size*)

"The likenesses are often so accurate that I surprise myself."

The typewriter process
"I use a manual typewriter and a carbon duplicating cash-register roll. I start with the hair, and then move side to side and outwards to the ears, down to the nose and mouth, finishing with the shoulders, always working downwards and across and backwards and forwards, always looking and looking; it is quite an intense process. Most of this portrait work takes place in public, at events and art fairs where I find myself plonked right in the middle of the gallery under a spotlight. I am becoming like a performance artist, and I am not a particularly outgoing person; I have to throw myself in at the deep end. My portraits are really popular and there are usually queues of people and kids hanging on to my arms, and I really do have to concentrate. It takes me about three minutes for a child and five minutes for an adult, and I have done twelve portraits in an hour. I often think this is ridiculous to put myself through all this, but typewriter portraits are what I do at the moment. But I do not want to overdo them; I want to keep them special. I am sure it will all move on to something else."

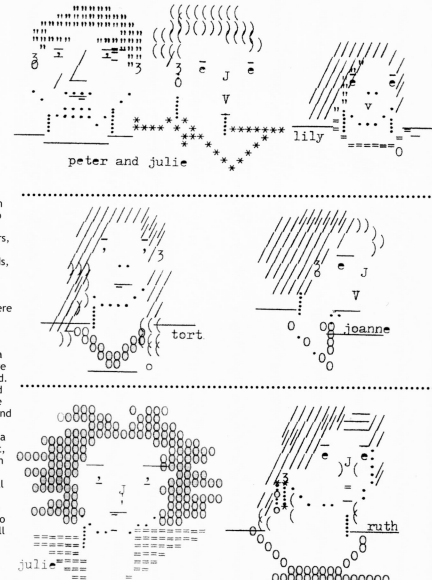

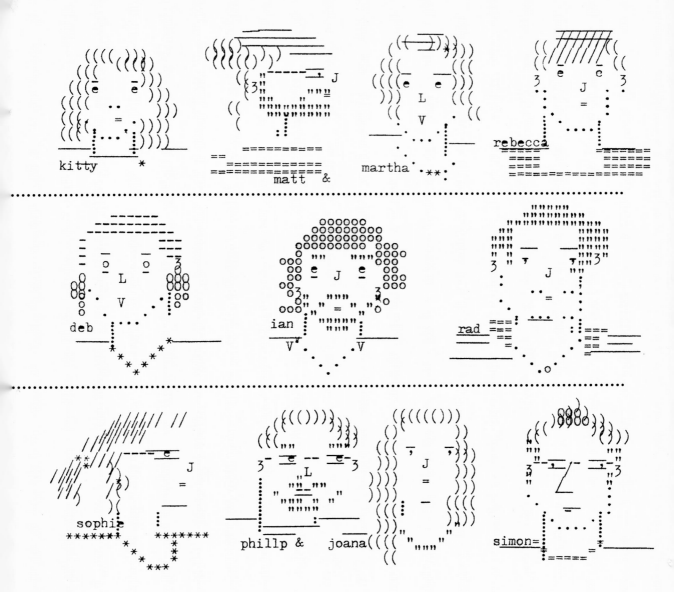

kitty

matt &

martha

rebecca

deb

ian

rad

sophie

phillp &   joana

simon

See also
Visual reference, page 178
Photographic sources, page 179

The creative manipulation of photographs has been used by artists since the beginning of photography, and altering and editing images originating in the camera is a well-known artistic genre. Cameras, in their own right, play a valuable role as reference and recording tools for artists (see page 178). However, in this digital age, when used in conjunction with computers, the boundaries of possibility for creative expression can be pushed even further.

Creating a graphic portrait in the computer is a simple process of deconstruction and reconstruction.

In the first stage the original photographic image is taken into the computer either by scanning it from a print, or exporting it directly from a camera card. In the examples shown here and on the following pages the image is created by using an application known as *Photoshop*, a popular software programme familiar to most PC and Mac users.

Once the image is in the computer, after a bit of tidying up – the original may need squaring up in frame and may require a little retouching – the simple process of creating a graphic portrait begins.

The scan is enlarged on the screen to a workable size and the face outline is traced by using the pen tool.

This is the first step in the process of deconstruction, and this selected face area is saved as a separate layer. A flesh colour, close to the original, is now infilled to the face area and, using the appropriate tool, the crisp, dark graphic outline is drawn around the face.

The process is repeated and a new layer is made for the neck with similar colour and outline. Next the hair is defined on yet another layer; it is important to capture the hair correctly as this sits on the top layer and frames the face. The clothes and the background are treated in a similar manner.

Finally, the features are applied, starting with the bottom lip and the nose. The eyes are usually bold dots, with a white highlight.

The process creates amazing likenesses and is very popular with sitters.

Computer generated portraits
Taking an original photograph, either from a digital camera card, or from a flat print, the artist manipulates the image into a graphic rendition of the original. There are countless effects that can be created with the aid of a computer. In this case, a simple flat-colour graphic effect, similar to that of a silk-screen print, has been extracted from the original photographic image.

Team Portraits
"Snappysnaps"
Bristol and Newbury, 2007
Digital prints on paper

Team Portraits
(*Across the spread from left to right*)
Cecile, Mark I, Mark H, Dean, Nat,
Mark, Amelie, Fee, Nicci, Sarah,
Susi, Teresa

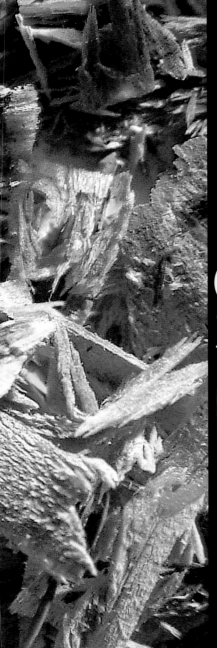

general
reference

You can make simple "flick" books by photocopying the images opposite and on the following pages. Follow the instructions to make little animated sequences of a rotating head and a closing eye.

Rotating head
For the turning head sequence assemble images 1-42 in running order.

Equipment
You will need the following: Photocopier or scanner, scalpel or craft knife, cutting mat (optional, a scrap of thick card will do equally well), steel rule or straight edge, heavyweight paper (about 150gsm, or 90lb, suitable for photocopiers or printers), glue (white PVA or stick glue is good), masking tape or similar tape.

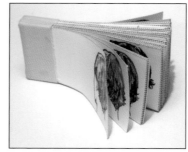

Opening eye
For the opening eye sequence assemble images 1-63 in running order.

Instructions

1. Scan or copy the images. Place the book on the bed of the scanner or photocopier; you may need to weight it down to make good contact with the surface to avoid distortion.

2. Cut the images out following the dotted lines. Each dotted area represents an individual page.

3. You can duplicate more copies than printed here. This will give you a fatter flick book and a longer animated sequence.

4. Arrange the pages in correct numerical sequence.

5. Glue the edges of the pages at the spine. Secure the book-block with tape.

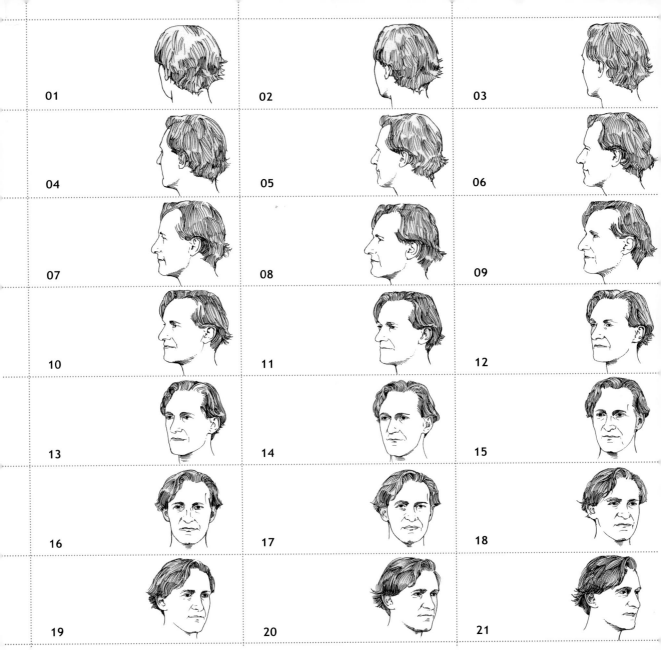

01 02 03
04 05 06
07 08 09
10 11 12
13 14 15
16 17 18
19 20 21

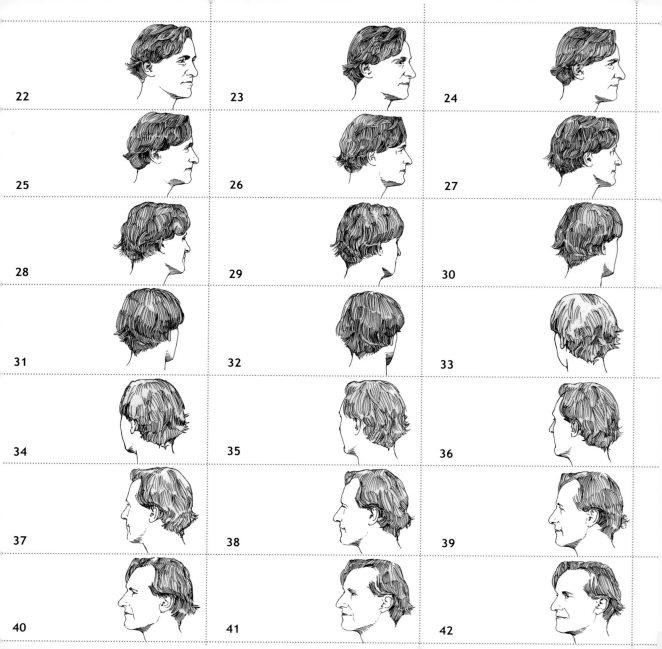

22

23

24

25

26

27

28

29

30

31

32

33

34

35

36

37

38

39

40

41

42

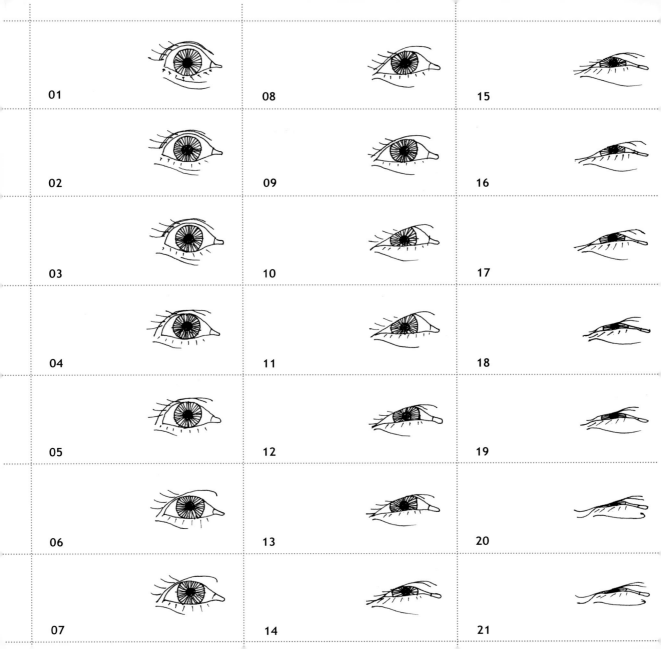

01

02

03

04

05

06

07

08

09

10

11

12

13

14

15

16

17

18

19

20

21

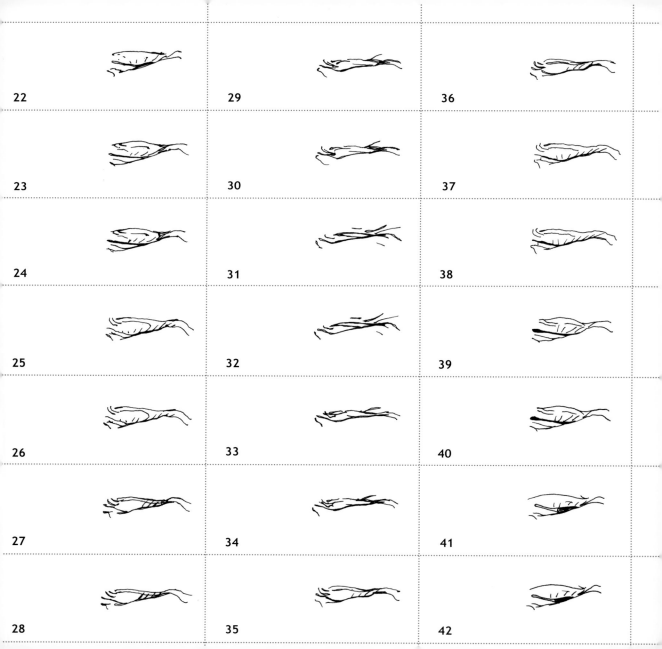

22

23

24

25

26

27

28

29

30

31

32

33

34

35

36

37

38

39

40

41

42

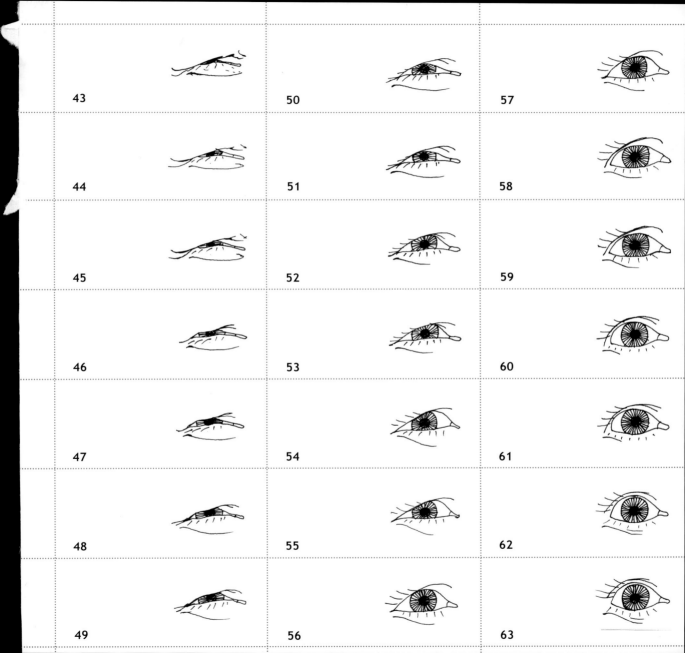

43

44

45

46

47

48

49

50

51

52

53

54

55

56

57

58

59

60

61

62

63

Broken colour

The details of paintings by Vincent van Gogh (*below*) and John Reay (*right*) are examples of flesh treatment where colour is conveyed by a technique known as "optical mixing". Colours are laid side by side to convey an overall colour sense.

There is a generalization about the colour of flesh – that it is either, black, white, brown, or pink, or shades thereof. This misconception is supported by the fact that many artists' colour manufacturers market colours denoted as "flesh tints", or "portrait colours", which are usually of a pink hue. These colours are perfectly good in their own right and have a place in the artist's palette. However, close study of the human face, no matter what the complexion of the sitter, shows that flesh colour is made up of a myriad of colours and tones. These are all combined with and influenced by light and shade, and colour reflection from the ambient surroundings. Colour experimentation and close observation of the sitter pay dividends when tackling the problem of human skin colour. It is worth looking closely at how other painters have dealt with the issue of portraying skin tones too.

Keep a sketchbook to experiment and play with colour.

Manufactured colours
Many artists' colour manufacturers describe colours in general terms of "flesh tints", or "portrait colours".

Keeping notes
Experimenting with colour, combined with careful observation of the subject, is a way of moving towards accuracy and understanding when depicting human skin colour. Also study how other artists have variously portrayed human flesh.

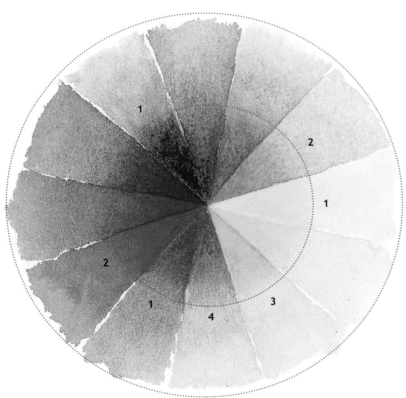

The colour wheel
One of the most important "tools" for the artist is the colour wheel (*left*). It is an arrangement of the primary colours (red, yellow, and blue) and secondary colours (orange, green, and violet) from which all others – including greys and browns – are mixed.

Primary colours
The primary colours are equidistant on the colour wheel. A primary colour is one that cannot be made by mixing other colours.

Secondary colours
A secondary colour is obtained by mixing two primaries. Thus, blue and yellow make green, red and yellow make orange, and red and blue make violet.

Tertiary colours
A tertiary colour is made by mixing a primary colour with the secondary next to it on the colour wheel. If you combine red with its neighbour to the right (orange), you get red-orange; if you combine red with its neighbour to the left (violet), you get red-violet.

| **1.**
Primary colours
Blue, yellow, and red. | **2.**
Complementary colours
Colours opposite to one another on the colour wheel. | **3.**
Secondary colours
Obtained by mixing two primaries. | **4.**
Tertiary colours
Made by mixing a primary with the secondary next to it. |

Complementary colours
The colours opposite one another on the wheel are known as complementary colours.

General reference
Colour | 177

Complementary colours
The colours opposite one another on the wheel are contrasting partners, and are called complementary colours. There are three main pairs, each consisting of one primary colour and the secondary composed of the other two primaries. Thus, red is the complementary of green, blue of orange, and yellow of violet. These relationships extend to pairs of secondary colours, so that red-orange is complementary to blue-green, blue-violet to yellow-orange, and so on.

Warm and cool colours
All colours have familiar associations. Reds and yellows conjure up sunlight and warmth; and we connect blues and greens with the coolness of water, foliage, and shadow. Taking advantage of these associations allows you to create a distinct atmosphere or mood in your painting.

Neutral colours
Pure neutrals are mixtures of black and white, and are neither warm nor cool. However, most so-called neutrals are the result of mixing two or more colours, and therefore have a temperature bias depending on the proportions of the colours used in the mix.

Tones
Tone refers to the relative lightness or darkness of a colour. Some colours are by their nature lighter in tone than others. For instance, cerulean blue is light in tone, while Prussian is darker.

Tints
When white is added to a colour to lighten it, the resulting mix is referred to as a tint of that colour.

Shades
Shades are darker tones of a colour, achieved by adding black.

Harmonious colours
There is a harmonious relationship between colours that lie on the same section of the colour wheel. The closest relationships are between shades of one colour, or between a primary colour and a secondary that contains that primary, such as blue and blue-green or blue and blue-violet.

Hues
Hue is another word for colour, and refers to the generalized colour of an object. The term is used to describe close or similar colours: for example, cadmium red and alizarin crimson are close in hue.

Colour mixing
Knowing how to mix and use colours is crucial to the success of your painting. Yet many artists find selecting, mixing, and using colours to be a bewildering process. Understanding the basic principles of colour theory and knowing how to apply colours in practice will help boost your confidence. Experiment in your sketchbook; in theory, by mixing the primary colours in varying proportions, you can produce every other colour known.

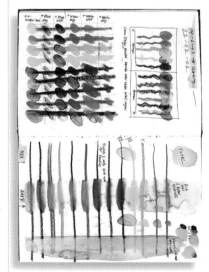

Photographs are acceptable as artists' reference material, provided you use them as you would use sketches, and not as an end in themselves.

When to use photographs

Photographs are especially useful in situations when there is very little time for drawing – such as when working with an active child, for example. Always try to make a few sketches at the same time, to augment your photographic record, and take as many shots as possible from different viewpoints during the session.

Building a reference library

If you are interested in portraiture, build a reference library, comprising your own photographs, with cuttings from newspapers, magazines, and brochures. This will provide you with a rich variety of angles, lighting conditions, and interesting characters to draw upon.

See also
Recording features and details,
pages 62-63
Model reference, pages 104-105

General reference
Photographic sources | 179

Wide-angle lens

**Beware of
lens distortion**
When taking your
own reference
photographs, do
not forget that
camera lenses
distort,
particularly wide-
angle lenses. This
is the type of lens
often used in
digital cameras,
and you will need
to adjust your
drawings
accordingly.

Standard lens

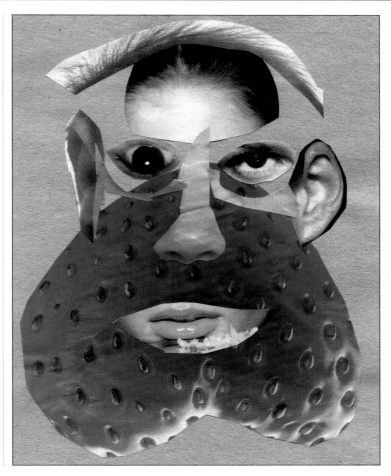

Collage
Assembling and
creating
artworks from
printed
ephemera and
found materials
is a time-
honoured
artistic
technique much
favoured by the
Surrealist
movement of
the mid-20th
century. Here
the artist has
followed the
tradition and
made this
unnerving
collage of the
human face
using a variety
of photographic
images cut from
magazine
advertisements.

Throughout this book you will have seen references to a variety of materials, some of which may be new to you. The following is a glossary of useful information that relates to many of the popular art materials.

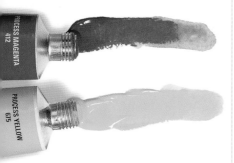

Oil paint
Oil paint is the traditional medium of choice for artists; it is slow to dry and is infinitely workable. It consists of dry pigments ground in a natural drying oil such as linseed, or a semi-drying oil such as safflower or poppy. It requires the use of solvents, such as turpentine or white spirit, to dilute the paint and clean up.

Graphite pencil
The common "lead" pencil is available in a variety of qualities and price ranges. The graphite core (lead) is encased in wood and graded, from softness to hardness: 9B is very, very soft, and 9H is very, very hard. HB is the middle-grade, everyday, writing pencil. The H-graded pencils are mostly used for technical work. For freehand drawing work, choose a pencil from around the 2B mark.

Graphite sticks
These are made of high-grade compressed and bonded graphite formed into thick, chunky sticks. They glide smoothly across the surface of the paper, lending themselves to bold, expressive drawing. You can vary the marks by using the point, the flattened edge of the point, or the side of the graphite stick.

Charcoal
Charcoal drawing sticks, made from charred willow or vine twigs, are available in three grades: soft, medium, and hard. Soft charcoal, which is very powdery, is ideal for blending and smudging; the harder varieties are better for linear drawing. When made into pencils, charcoal is cleaner to use and is more controllable.

Charcoal pencils
Charcoal pencils are thin sticks of compressed charcoal encased in wood. This means they are cleaner to handle and easier to control than stick charcoal. They have a slightly harder texture and only the point can be used, so they are not ideal for creating large areas of tone. Charcoal pencils come in hard, medium, and soft grades. The tips of the pencils can be sharpened, like graphite pencils.

Coloured pencil
This is a generic term for all pencils with a coloured core. There is an enormous variety of colours and qualities available. They also vary in softness and hardness, but, unlike graphite pencils, this is seldom indicated on the packet.

Watercolour pencil
As above, except water-soluble and capable of creating a variety of "painterly" effects, by either wetting the tip of the pencil or working on dampened paper.

Conté crayon
Often known as Conté pencil, this is available in pencil or chalk-stick form. Originally a proprietary name, Conté has become a generic term for a synthetic chalk-like medium, akin to a soft pastel or refined charcoal. It is available in black, red, brown,

and white, but is best known in its red form. Conté is a traditional and well-loved drawing medium.

Pastels
Pastels are drawing sticks made by blending coloured pigments with chalk or clay and bound with gum. This type of pastel is relatively soft and brittle. Oil pastels, made using an animal-fat binder, are stronger and less crumbly. Both types of pastels are available in a very wide range of colours and tones, as are pastel pencils.

Drawing ink
There is a variety of inks available, from water-soluble, writing (fountain-pen or calligraphy) inks to thick, permanent, and waterproof drawing inks. India ink is a traditional drawing ink; it is waterproof and very dense, drying with an interesting, shiny surface. Drawing inks are available in many colours, and they can be thinned down with distilled water for creating washes.

Steel-nib (dip) pen
The old-fashioned, dip-in-the-inkwell pen is a worthy and versatile drawing instrument. You may want to experiment with nibs for thickness and flexibility, but just a single nib can make a variety of line widths as you alter the pressure on the pen.

Sketching pens
Although they resemble ordinary fountain pens in appearance, sketching pens have flexible nibs (designed specifically for drawing) that deliver ink smoothly from prefilled ink cartridges to the drawing paper.

Fibre-tip pen
Fibre-tip pens come in a variety of tip thicknesses. The ink flows well and smoothly, and fibre-tips are useful for creating drybrush effects when the ink runs out.

Erasers
You can rub out unwanted pencil marks and drawings with an eraser, and you can also use it to create highlights by lifting out areas of tone. Plastic or putty erasers are the best to use; India rubber tends to smear pencil work, and it can damage the paper surface. Putty erasers are very malleable; you can break off smaller pieces and manipulate them into any shape, including a point for lifting out.

Paper stump
A paper stump, sometimes known as a tortillon or torchon, is a strip of paper twisted into a narrow cone. The point is used to blend and soften charcoal or pastel.

Watercolour paint
A popular paint for amateurs, yet it can be notoriously difficult to use. Watercolour paints are made of very finely ground pigments bound with gum-arabic solution. The gum enables the paint to be diluted with water, allowing thin transparent washes of colour that will adhere to the paper. Watercolours are available in artists' or students' quality.

Acrylic paint
A paint that contains acrylic resin obtained by the polymerization of acrylic acid. It is a popular painting medium with similar properties to both oil and watercolour paints. Acrylic colour can be mixed with water, but dries very quickly to a hard, plasticky film.

Gouache
The matt, chalky appearance of gouache and its opacity when dry are very different qualities from the pureness and transparency of watercolour – but both media use similar equipment and techniques. The best-quality gouache paints are pure and intense, and they create clean mixes. The less expensive gouache colours, also known as poster paints, often contain an inert white pigment, such as chalk, in order to impart smoothness and opacity. Gouache paint is sold in tubes, pots, and bottles.

The subject of art has its own special terminology. Listed below are some terms that you may come across.

Abstract
A style of painting where the figurative depiction of subjects has been entirely discarded (or almost entirely discarded, **Semi-abstract**), and represented by non-representational forms, shapes, patterns, and marks. (*See also Expressionism, Marks*)

Academy
(1) A distinguished institution for the study of art. **(2) (Academic)** Relating to a traditional or conservative style of art. (*See also Classical/Classic*)

Alla prima
(Italian for "at the first") A technique in which the final surface of a painting is completed in one sitting, without underpainting. Also known as "au premier coup".

Atmospheric perspective
The illusion of depth created by using desaturated colours and relatively pale tones in the background of a painting. Also known as "aerial perspective". (*See also Perspective*)

Balance
In a work of art, the overall distribution of forms and colour to produce a harmonious whole.

Blending
Smoothing the edges of two colours together so that they have a smooth gradation where they meet.

Blind drawing
An exercise where a drawing is made with the eyes closed or made without looking at the surface on which one is drawing.

Body colour
Opaque paint, usually water-soluble, such as gouache, that has the covering power to obliterate underlying colour. "Body colour" also refers to watercolour mixed with white paint, and a pigment's density.

Calvaria
The rounded "cap" of the skull.

Caricature
A portrait in which the characteristics and features are exaggerated, often for comic or grotesque effect.

Chiaroscuro
(Italian for "light-dark"). Particularly associated with oil painting, this term is used to describe the effect of light and shade in a painting or drawing, especially where strong tonal contrasts are used.

Classical/Classic
Highly figurative and representational paintings that follow the general precepts of ancient Greek and Roman art, or were genuinely painted in that period, up until the beginning of the 19th century. (*See also Academic, Master paintings*)

Collage
Artwork composed of pieces of paper, fabric or other items pasted onto a support or drawing.

Composition
The arrangement of subject matter within a picture. It involves the planning of colour, use of light and shade, and considered placing of the various elements.

Contre-jour
(French for "against daylight") A painting or drawing where the light source is behind the subject.

Contour
A two-dimensional outline representing or bounding the three-dimensional shapes or forms of a subject in order to understand or describe its volume.

Cranium
The whole skull, except for the jawbone. (*See also Mandible*)

Crosshatching
Close parallel lines that crisscross each other at angles, used to model and indicate tone. (*See also Hatching*)

Diluent
A liquid, such as turpentine, used to dilute oil paint. The diluent for water-based media is water.

Dot and stipple
A drawing technique where tone and shade are created with "pin-point" dots. Usually executed in pen and ink.

Dry brush
A technique for applying the minimum of paint by lightly stroking a barely loaded dryish brush across the surface of a painting or support.

Etching
A print or impression produced from a drawing scratched into a metal plate. Areas of tone and texture may be added by dipping the plate into acid. The plate is then inked and passed through a printing or etching press.

Expression
In portraiture, the mood of the subject conveyed by the features, particularly by the eyes and mouth.

Expressionism
A style of art that is usually spontaneous and characterized by intense feelings. (*See also Abstract*)

Features
In portraiture, the elements – ears, nose, eyes, hair, etc - that contribute to the unique facial appearance of an individual.

Filbert
A flat-bristle brush with a rounded end similar in shape to a fingernail.

Format
The size and proportions of a painting or drawing. The two most popular are traditionally landscape and portrait, but square and round are frequently also chosen. (*See also Landscape format, Portrait format*)

Fugitive colours
Pigment or dye colours that fade when exposed to light.

Genre
A category or type of painting classified by its subject matter – such as still life, landscape, portrait, etc. The term is also applied to scenes depicting domestic life.

Gesso
A sealant mixture, usually composed of whiting and glue size, applied as a primer and ground for rigid painting supports.

Glaze
In painting, a transparent or semi-transparent colour laid over another colour to modify or intensify it.

Graffiti
The defacement of walls and buildings with sprayed, scratched, and painted images and texts. (*See also Sgraffito*)

Ground colour
A specially prepared painting surface. Dilute or broken colour is applied to a primed canvas or other support in order to reduce the glare from a white surface.

Hatching
A technique for indicating tone and suggesting light and shade in a drawing or painting, using closely set parallel lines. (*See also Crosshatching*)

Highlight
The lightest areas of tone within a painting or drawing, generally describing reflected light.

Hue
The name of a colour – blue, red, yellow, etc – irrespective of its tone or intensity.

Impasto
A technique of applying paint thickly with a brush or painting knife, or by hand, to create a textured surface. Also the term for the results of this technique.

Intuitive
An approach to artwork that is natural and unpremeditated.

Key
Used to describe the prevailing tone of a painting: a predominantly light painting has a high key, a dark one has a low key.

Landscape format
A painting or drawing where the height is narrower than the width. This is the traditional shape for painting a landscape or vista. (*See also Portrait format*)

Linework
The linear elements of a picture, produced by drawing with a pen, pencil or paintbrush.

Mandible
The movable lower jaw, including the chin. (*See also Cranium*)

Marks
The physical and visual characteristics of applied brushstrokes, drawn lines etc. They are the fundamental components of an artwork. (*See also Abstract, Expressionism*)

Master paintings (Old Masters)
Paintings associated with a period of history in Europe that lasted from approximately the 14th to the 16th centuries. The central period known as the Renaissance saw a rebirth and flowering of the arts, science and culture, much of which was associated with the great master Italian painters in the south, and the Dutch and Flemish painters of the north. (*See also Academy, Classic*)

Medium
This term has two distinct meanings: **(1)** an additive (plural **mediums**) mixed with paint to modify characteristics such as flow, gloss, or texture; **(2)** the material (plural **media**) chosen by an artist for working, such as paint, ink, pencil, pastel, etc.

Mixed media
In drawing and painting this refers to the use of different media in the same picture (for example, ink, watercolour wash, and wax crayon), or to the use of a combination of supports (for example, newspaper and cardboard).

Modernism
A period in the history of art, generally from the early to the late 20th century, encompassing movements such as Cubism, Futurism. Also refers to "new" ways of expression in art, design, and architecture associated with the early to mid 20th century.

Monochrome
An image made using different tones of black and white or of one colour.

Orbital opening
Openings in the skull which contain the eyes and associated muscles.

Outsider art
Work executed by untrained artists. (*See also Primitive art*)

Palette
The tray or surface on which colours are arranged and mixed. Palette also refers to the particular selection of colours that an artist chooses.

Performance art
An art form where a range of physical actions, environmental conditions, and situations become the artwork. Similar to a theatrical performance.

Perspective
A method of establishing space and depth in a painting or drawing by describing objects in the distance as smaller than those in the foreground. Parallel lines converge to a vanishing point as they meet the horizon (eye level). Also called linear perspective. (*See also Atmospheric perspective*)

Philtrum
The vertical groove in the upper lip, allowing humans to express a large range of lip motions.

Phizog
Referring to the human face. From "phiz" (abbreviation of physiognomy), old British slang for face.

Physiognomy
A person's facial features or expression comprising the skin, eyes, and hair etc, which bring character to each individual.

Pictograph
A pictorial symbol for a word or phrase (also known as a pictogram).

Picture plane
The vertical surface or area of a drawing or painting in which the composition is placed.

Plein air
(French for "open air"). Term describing paintings done outdoors, directly from the subject.

Portrait format
A painting or drawing where the height is taller than the width. As its name suggests, this is the traditional shape for painting a portrait. (*See also Format, Landscape format*)

Primer
Applied to a layer of size or directly to a support, a primer acts as a barrier between paint and support. It also provides a surface suitable for receiving paint. (*See also Ground colour*)

Primitive art
Originally referring to prehistoric or tribal art, work executed by intuitive or untrained artists. (*See also Outsider art*)

Profile
In portraiture, used to describe a side view of the human head.

Proportion
The relationship of one part to the whole or to other parts; for example, the relation of each component of the human figure to the figure itself or to the painting as a whole.

Recession
In art, this describes the effect of making objects appear to recede into the distance by the use of atmospheric perspective and colour.

Resolution
(1) The final stages in finishing a painting. (2) The sharpness or "focus" of an image.

Semiotics
The study of signs and symbols, and their use and interpretation.

Sgraffito
(Italian for "scratched off") Scoring into a layer of colour with a sharp instrument, to reveal either the ground colour or a layer of colour beneath.

Shade
(1) A colour that has been darkened by the addition of black. (2) An area of an artwork that is in shadow.

Study
A detailed drawing or painting made of one or more parts of a final composition, but not of the whole.

Support
A surface used for painting or drawing, such as canvas, board, paper, etc.

Tint
Term for a colour lightened with white. Also, in a mixture of colours, the tint is the dominant colour.

Tone
The lightness or darkness of an element in a drawing or painting, irrespective of its local colour. Also called value. The term "tonal value" is also used to describe the relation of one tone to another.

Tooth
(Also known as Grain).The surface texture, ranging from coarse to fine, of paper, canvas etc.

Torso
The trunk of the human body. In sculpture, a statue without head or limbs.

Trompe l'œil
(French for "deception of the eye") A painting that is so realistic that the viewer is fooled into thinking the objects or a scene are real.

Underpainting
An early stage of a painting (sometimes in monochrome) used to establish the composition, overall tone, and colour balance. Also known as "laying in". (*See also Composition, Monochrome*)

Value
The term "tonal value" refers to the relative degree of lightness or darkness of any colour. (*See also Tone*)

Viewfinder
A frame used by artists, usually cardboard, to help concentrate the eye by isolating the subject from extraneous background.

Viewpoint
In portraiture, the position of the head in relation to the onlooker: three-quarter, full-face etc. (*See also Profile*)

Volume
The space filled by an element or object in a drawing or painting. (*See also Contour*)

Wash
A thin, usually broadly applied, layer of transparent or heavily diluted paint or ink.

Wet in wet
A watercolour technique for mixing two or more colour washes on a support before the washes have had time to dry.

Wet on dry
A watercolour technique for applying a wet wash over another that has already dried on the paper.

In the portrait studio, pages 76-107

(at Dunsay Manor Studio, Dorset, UK, 2007)

Artist:
Toby Wiggins
Many thanks to Toby Wiggins and his portrait model Catherine Smith for their generous contributions.

Photographer:
All studio photography, location and model reference photography is by **Ben Jennings**

Artist's Models
Many thanks to the life models of The Art House and Marina Studio who gave generously of their time and appear on pages 15, 19, 20, 38, and 56.

© The copyright in the images reproduced in **Face Parts**, *a visual source book for depicting the human face*, remains with the individual artists, illustrators, photographers, organizations and/or their successors, and is reproduced here by prior arrangement and/or courtesy of.

Some of the material in this book first appeared in The Ways of Drawing series, Inklink © 1993, © 1994, and in Vincent, A Complete Self Portrait, Inklink © 1994.

Further titles by Simon Jennings:
Body Parts, 2007 (also published by Mitchell Beazley)
The Artist's Colour Manual, 2003
Art Class, 1999
The Artist's Manual, 1995

The editors and producers of **Face Parts**, *a visual source book for depicting the human face*, have attempted to contact all copyright holders of material published in this book. In case of any errors or omissions, please contact the publishers for rectification in subsequent editions.

Picture credits/image sources
All images are identified in the following listing by **Section** and page number (p.)

Behind the face
Ben Jennings (p.10, 11, 13); Val Wiffen (p.11, 13, 15); Simon Jennings (p.14); *The Household Physician*, Gresham Publishing Co. 1932 (p.14); *The Concise Home Doctor*, The Amalgamated Press *c.*1910 (p.15).

Starting points
Ben Jennings (p.16-17, 19, 20, 22, 23, 24, 28); Simon Jennings (p.21, 25, 33); Val Wiffen (p.20, 26, 27, 29, 32, 33); Victor Ambrus (p.21); Nick Hyams (p.21); Naomi Russell (p.24); Roger Coleman (p.28, 30, 31).

Features and details:
Eyes
Ben Jennings (p.34-35, 36, 38, 39, 40, 42); Michael Woods (p.38, 39); *The Household Physician*, Gresham Publishing Co. 1932 (p.37); Roger Coleman (p.40, 41, 42, 43, 44, 45, 46, 47, 48, 49, 50, 52, 53); Val Wiffen (p.44, 45, 47, 49, 51).

Features and details:
Ears, mouths, noses & hair
Ben Jennings (p.54-55, 56); *The Household Physician*, Gresham Publishing Co. 1932 (p.61, 65, 67); Roger Coleman (p.57, 64, 68);

Val Wiffen (p.60-61, 64, 65, 66, 67, 68-69, 70-71, 74-75); George Underwood (p.59); Simon Jennings (p.58, 59, 69); British Museum, (p.64); Alain Platel/The Shout (p.64).

In the portrait studio
Toby Wiggins (p.76-107, *all artwork*); Ben Jennings (p.76-107, *all photographs*); National Gallery, London (p.106-7).

Self portraits
Desmond Haughton (p.110); National Gallery, London (p.111); Vincent van Gogh (p.112-119, *all artwork images courtesy Inklink/Bernard Denvir, from Vincent, A Complete Self Portrait, Inklink © 1994*); Uffizi Gallery, Florence (p.120); Simon Jennings (p.108-109, 121, 122, 123).

Art directions
Gemma Anderson (p.126-127); George Underwood (p.128-129, 130-131); Simon Jennings (p.132, 133, 134-135, 136-137, 138-139, 140, 142-143, 144-145, 146-147, 150-151, 152-153); Henry Dreyfuss, *Symbol Sourcebook/An International Dictionary of Symbols* (p. 140-141); Nadine Faye James (p.158-159, 160-161); Snappy-Snaps, Bristol (p.162-165, *all computer generated portraits*); *The Art and Craft of Hairdressing*, New Era Publishing *c.*1930 (p.154-157, *all images*); University of Southern California Archives (p.149). *(William Tuttle Obituary by Ronald Bergan, The Guardian Newspaper [UK] 23.8.2007).*

General reference
Simon Jennings (p.166-167, 168, 179); John Reay (p.175); Ink-Archive (p.169-173); Ben Jennings (p.174-179 *all photos*).

Face Parts
Created by Simon Jennings

First published in Great Britain in
2008 by Mitchell Beazley,
an imprint of Octopus Publishing
Group Limited,
2-4 Heron Quays, London E14
An Hachette Livre UK Company
www.mitchell-beazley.com

Conceived, edited, and designed by
Simon Jennings at Auburn Studio.

Text editor Geraldine Christy
Photography Ben Jennings

Some of the material in this book was first published in
The Ways of Drawing series
(Running Press/Cassell-Studio Vista 1994),
and in *Vincent, A Complete, Self Portrait*
(Running Press/Pavilion 1994)

Also available in the series:
Body Parts,
*a practical source book
for drawing the human form*
ISBN: 9 781 84533 3188

Face Parts, *a visual source
book for depicting the human
face,* has been made possible
by the goodwill and
cooperation of many people,
and represents the work of
several hands and minds. The
editors and producers are
grateful to the following artists
for their valuable contributions
to this book.

Contributing artists

Victor Ambrus
Gemma Anderson
Roger Coleman
Desmond Haughton
Nick Hyams
Nadine Faye James
Simon Jennings
John Reay
George Underwood
Val Wiffen
Toby Wiggins
Michael Woods

*All artwork © The Artists (see page 191).
Some artwork first appeared in The Ways of
Drawing series, Inklink © 1993, © 1994.*

Consultants

Toby Wiggins,
In the portrait studio section.

Carolynn Cooke,
Art education consultant.

A CIP catalogue record
for this book is available from
the British Library.

ISBN: 978 1 84533 4130

To order this book as a gift or an
incentive contact
Mitchell Beazley on 020 7531 8484.

Set in: Trebuchet MS
Colour reproduction by
United Graphics
Printed and bound by
Toppan, China